SCIENTIFIC JOURNALS
IN THE UNITED STATES

**PUBLICATIONS IN
THE INFORMATION SCIENCES**

Rita G. Lerner, Consulting Editor

TWO CENTURIES OF FEDERAL INFORMATION/*Burton W. Adkinson*
INDUSTRIAL INFORMATION SYSTEMS: A Manual for Higher Managements
and Their Information Officer/Librarian Associates/*Eugene B. Jackson
and Ruth L. Jackson*
LIBRARY CONSERVATION: Preservation in Perspective/*John P. Baker and
Marguerite C. Soroka*
SCIENTIFIC JOURNALS IN THE UNITED STATES: Their Production Use,
and Economics/*Donald W. King, Dennis D. McDonald, and
Nancy K. Roderer*

SCIENTIFIC JOURNALS IN THE UNITED STATES

Their Production, Use, and Economics

Donald W. King
Dennis D. McDonald
Nancy K. Roderer
King Research, Inc.

with contributions by
Patricia M. Dowd
Charles G. Schueller
Barbara L. Wood
Mary K. Yates

Hutchinson Ross Publishing Company

Stroudsburg, Pennsylvania

All of the material incorporated in this work was developed
with the financial support of National Science Foundation
grant number NSF C-DS175-06942 (formerly C878).
However, any opinions, findings, conclusions or
recommendations expressed herein are those of the
authors and do not necessarily reflect the views of the
National Science Foundation.

83   82   81   1   2   3   4   5
Manufactured in the United States of America.

Library of Congress Cataloging in Publication Data
King, Donald Ward, 1932-
    Scientific journals in the United States.
    Includes index.
    1. Communication in science—United States.
2. Communication of technical information—United
States.   3. Science—United States—Periodicals.
4. Engineering—United States—Periodicals.
I. McDonald, Dennis D., joint author.   II. Roderer,
Nancy K., joint author.   III. Title.
Q223.K539   070.4′495   80-25945
ISBN 0-87933-380-4

Distributed world wide by Academic Press,
a subsidiary of Harcourt Brace Jovanovich,
Publishers.

Contents

Preface

Chapter 1   Scientific and Technical Information Transfer
    Through Journals   1
The Importance of Scientific and Technical Information
    Transfer   1
The Scientific and Technical Information Transfer
    Environment   3
The History of Scientific and Technical Journals   7
The Scientific and Technical Journal System   10
Chapter 2   Summary Analysis   21
The Scientific and Technical Communication
    Environment   21
The Growth of the Scientific and Technical Literature   21
The Scientific and Technical Journal System   28
Indicators for the Scientific and Technical Journal
    System   28
Some Observed Weaknesses in the Journal System   37
New Technology for the Journal System   41
Chapter 3   Authorship of Scientific and Technical Journals   49
Description of the Scientific and Technical Community   49
Description of Journal Authors   55
Number of Journal Articles Published   59
Authors' Selection of Journals to Which to Submit
    Articles   62
Journal Authorship Activities   65
A Model of Manuscript Flow   67
Special Considerations in the Author and Publisher
    Interface   72
Cost of Journal Authorship   75
Contribution of Federal Funding to Authorship   78
Chapter 4   Publishing Scientific and Technical Journals   81
Size and Growth of Journal Publishing   81

vi   Contents

General Characteristics of U.S. Scientific and Technical
Periodical Publishing   89
Economic Characteristics of U.S. Scientific and
Technical Journals   94
A Cost Model for Journal Publishing   109
Federal Contribution to Publishing   125
Chapter 5   Libraries and Secondary Services   129
Library Activities   129
Library Cost Associated with Scientific and Technical
    Journals   142
Federal Contribution to Libraries   146
Secondary Services Activities   146
Federal Contribution to Secondary Services   157
Chapter 6   Use of Scientific and Technical Journals   161
Description of Scientific and Technical Journal Use   161
The Relation of Citation Analyses to Journal Use Data   170
Cost of Use of Scientific and Technical Journals   176
Federal Contribution Funding to Use of Scientific and
    Technical Journals   180
Chapter 7   The Flow of Information Through Scientific and
    Technical Journals   187
The Flow of Journals   187
The Impact of Time on Journal Flow   200
Communication Flow Between the United States and
    Other Countries   201
Chapter 8   Economics of the Scientific and Technical
    Journal System   207
An Economic Framework for Journal Systems Analysis   207
The Costs of Scientific and Technical Journals   215
Economic Interface Between Publishers and Users and
    Publishers and Libraries   229
The Social Benefit of Scientific and Technical Journals   238
An Assessment of Alternative Subscriptive Pricing
    Policies   243
Economic Interface Between Authors and Publishers   250
Federal Contribution to Scientific and Technical
    Communication   253
Chapter 9   Hypothetical Economic Analysis of the Journal
    System   265
An Individual's Subscription Decision   265
A Library's Subscription Decision   269
Effect of the New Copyright Law on Scientific and
    Technical Journals   276

Chapter 10 The Future of the Scientific and Technical
                  Journal System   285
   The Future of Authorship   285
   The Future of Publication   290
   The Future of Libraries and Secondary Organizations   297
   The Future of Journal Use   303
   The Future of Journal Transmission   305
   The Future Cost of the Journal System   308
   A Comprehensive Electronic Journal System   308
   Constraints on an Electronic Journal System   310

Index   315

Preface

This book presents the results of the final year of a three-year project, "Statistical Indicators of Scientific and Technical Communication." The project over these three years attempted to follow a research cycle in which research needs were formulated, models were developed, data were collected, and observations were analyzed and interpreted. The research cycle was repeated each year. Each cycle iteration was designed to reduce uncertainty in data or assumptions concerning them. The initial research needs were specified by the National Science Foundation:

> Numerous studies have focused on the transfer of research results from originators to users. But most of these investigations have been limited to one phase of information transfer or to one field of science or to one point in time. The need for a comprehensive and continuing information base for policy and planning is by no means new, but it becomes even more urgent as the volume of scientific and technical information rises beyond the reach of manual processing, while automation on a massive scale approaches the point of economic feasibility. In order to monitor and assess these developments a statistical system for gathering, organizing, and analyzing data relevant to all aspects of scientific and technical communication is indispensable. The objective of this project is to develop and initiate a system of statistical indicators of scientific and technical communication. Each indicator, as a time series, will trace out some aspect of the production of new scientific and technical information, of the transfer of that information to users, or of the resources needed to preserve it for future use and to retrieve it when needed. The indicators are intended for analysis and interpretation by planners and policymakers in Government and in the private sector and as data in modeling and simulation studies. Taken together they can be used to describe trends, anticipate new developments, and guide the evolution of the Nation's information resources toward better and more economical service.

From that statement of research needs, we formulated a conceptual model of scientific and technical communication that consisted of designating principal participants in the system and their roles, functions performed, products and services involved, and activities necessary to produce the products or operate the services. A constraint on the initial effort was that data employed in the study were limited to existing secondary sources; we could not conduct a survey to collect new observations. A substantial amount of secondary data was collected from previous studies and other data sources, and from them we derived a series of statistical indicators concerning the literature published in the United States. The indicators consisted of such measures as the amount of published literature, extent of distribution, average price, unit cost, and overall resource expenditures. Exhaustive data were collected back to 1960, and time-series analyses were made to project the measures from 1974 to 1980. The results of the first year's effort were published in three volumes.[1,2,3]

The second set of research needs involved validating the time-series models by collecting data from the year 1975 and testing the observations against projections. This resulted in some modifications of the time-series models, although the results were generally quite close, as one would expect after only one year. Also if was felt necessary to define the periodical literature more carefully. This redefinition resulted in a completely new analysis of the journal mode of communication. Another task was to extend the statistical indicators of the published literature to include worldwide data. All of these analyses were performed on secondary data except for an author survey. The final research needs included interest in indicators of the extent of numeric data base activity and amount of involvement of federal funding in scientific and technical communication. Primary data were collected by us to estimate numeric data base activity. Second, data were collected by NSF and analyzed and reported by us concerning federal funding in scientific and technical communication. During this year's effort, the cost models we employed in the analysis were also refined and some detailed economic analyses were presented. Results of the second year's effort were published in four volumes.[4,5,6,7]

In the meantime, members of our staff conducted a systems analysis of scientific and technical communication with particular attention to new technology (under NSF sponsorship). Emphasis was placed on electronic processes and on an electronic alternative to paper-based publishing. Much of the cost modeling developed in the statistical indicators study was used in this systems analysis. The conceptual model of system functions was refined, the cost model improved, and much further analysis done on use of scientific and technical journals based on secondary data. In this project and the second year's statistical indicators study, it became clear that some primary data collection was necessary to enhance the cost and flow models further. This conclusion was based on an analysis of the sensitivity of certain variables, the lack of precision of others, and a lack of consistency of still others. We felt the need for an in-depth study of at least one communication medium and decided that the journal was the most appropriate one for such thorough analysis. In our final effort, then, we developed primary data collected through surveys, including a journal user survey, a journal author survey, a journal publisher survey, and a data-base organization survey. Data were also collected from secondary sources for the year 1977 to update the statistical indicators. Time-series analyses were used to project new results to 1985.

Previous work did not show what portion of the federal government research funding was expended on scientific and technical communication. To estimate this proportion, we conducted a survey of federally funded research projects. These data, together with other results, yielded a more complete description of federal involvement in scientific and technical communication.

Analysis of the third year's study concentrated on the results of the five surveys and on tying together the entire three-year effort. Also the cost model was further refined and programmed for computer manipulation. Thus, one can incorporate into the programmed model such factors as alternate process variables (for example, equipment) or alternate cost elements (for example, equipment rates, personnel salaries, supply unit costs, or overhead costs). The model then calculates the direct effect of such alternatives on cost. The limitation here is that interactive effects must be introduced manually for each alternative.

This book serves as the report of the third year's effort on the statistical-indicators project. The first chapter provides a description of our conceptual model of the journal system of scientific and technical communication. The second chapter summarizes the results of analysis of the overall scientific and technical communication system in general and of the journal system in particular.

Generally this book is organized by the principal participants of the journal system. Thus, Chapters 3 through 6 are about authors, publishers, libraries and secondary services, and users. Within each of these four chapters, we broadly describe the functions related to each participant in terms of the total amount of activity involved and general trends since 1960 (when data are available). The trends are projected in most instances to 1985. Cost analysis is also given in each chapter and the appropriate portion of the cost model is discussed. Finally, the extent of federal fiscal involvement is discussed in each chapter.

Chapter 7 examines the flow of information through the journal system. Each principal path is described in terms of its flow and number of uses. Chapter 8 presents a detailed discussion of the economics of journal publishing, focusing on the economic principles of performance, effectiveness, benefit, and cost. Total costs of the journal system are presented and journal market, pricing policies, and the value of journals are described. Chapter 9 gives some hypothetical examples of the potential effect of interlibrary lending, networking, and the new copyright law.

The last chapter delves into the future of the journal system in the United States. In particular, trends in the system are forecast with the potential influence of innovations simulated in terms of their cost and effectiveness. The potential effect of technologies such as electronic processes, microform publishing, and telecommunications are explored. Also examined are the implications of library networks and resource sharing.

Throughout the statistical-indicators study, we have relied heavily on several sources of information. One of the most important sources of data concerning journal publishing has come from Professor Machlup and his colleagues.[8] These data were extremely useful in testing our publishing cost models and for establishing rates of change in number of subscriptions by type of subscribers over time. The work performed by Fry and White at Indiana University was also useful in cross-checking trends in growth rates, particularly involving libraries.[9] Research by Garvey and his associates helped set forth a framework within which we placed the journal system.[10] Other publications such as *Statistical Abstracts of the United States*,[11] the *Bowker Annual of Library and Book Trade Information*,[12] and *Ulrich's International Periodicals Directory*[13] were essential sources of secondary data.

In addition to the principal authors and other contributors, a number of other persons contributed substantially to this work. Helene Ebenfield of the National Science Foundation was involved in the project from its beginning to its completion. Her suggestions contributed a great deal. Consultants in earlier years included

F. W. Lancaster, Dr. F. Kertesz, K. W. Otten, W. A. Creager, and B. Goldwasser. Dr. Stephen J. Tauber assisted in editing and proofreading early drafts of this manuscript. A great deal of the analyses were performed by Leslie Gerran and Lisa King. Over the three years, in addition to the volumes referenced, we have presented more than fifty invited lectures, seminars, and presentations concerning this work. This has yielded a great deal of useful feedback to us in this project.

Referring back to the general research cycle that was used in this project, we consider much of what is presented here as identifying future research needs. We hope other information scientists will analyze these results, test them, confirm or refute our models and assumptions, and improve upon them in the future.

Donald W. King

REFERENCES

1. D. W. King, D. D. McDonald, N. K. Roderer, and B. L. Wood, *Statistical Indicators of Scientific and Technical Information Communication (1960–1980),* vol. 1: *A Summary Report* (Washington, D.C.: Government Printing Office, 1976).
2. D. W. King, F. W. Lancaster, D. D. McDonald, N. K. Roderer, and B. L. Wood, *Statistical Indicators of Scientific and Technical Information Communication (1960–1980),* vol. 2: *A Research Report* (Rockville, Maryland: King Research, 1976).
3. D. W. King, D. D. McDonald, and N. K. Roderer, *Statistical Indicators of Scientific and Technical Information Communication (1960–1980),* vol. 3: *A Data Appendix to Volume II* (Rockville, Maryland: King Research, 1976).
4. D. W. King, B. L. Wood, and C. G. Schueller, *A Chart Book of Indicators of Scientific and Technical Communication in the United States* (Rockville, Maryland: King Research, 1977).
5. D. W. King, D. D. McDonald, N. K. Roderer, C. G. Schell, C. G. Schueller, and B. L. Wood, *Statistical Indicators of Scientific and Technical Communication (1960–1980),* 2d ed. (Rockville, Maryland: King Research, 1977).
6. N. K. Roderer and C. G. Schell, *Statistical Indicators of Scientific and Technical Communication Worldwide* (Rockville, Maryland: King Research, 1977).
7. B. L. Wood, *Review of Scientific and Technical Numeric Data Base Activities* (Rockville, Maryland: King Research, 1977).

8. F. Machlup and K. W. Leeson, *Information Through the Printed Word: The Dissemination of Scholarly, Scientific and Intellectual Knowledge,* vol. 1: *Book Publishing;* vol. 2: *Journals;* vol. 3: *Libraries* (New York: Praeger, 1978).

9. B. M. Fry and H. S. White, *Publishers and Libraries: A Study of Scholarly and Research Journals* (Lexington, Mass.: Lexington Books, 1976).

10. William D. Garvey, *Communication: The Essence of Science* (Elmsford, New York: Pergamon Press, 1979).

11. U.S. Department of Commerce, Bureau of the Census, *Statistical Abstracts of the United States* (Washington, D.C.: Government Printing Office, 1960–1978).

12. *Bowker Annual of Library and Book Trade Information.* 4th–22d eds. (New York: Bowker, 1961–1977).

13. *Ulrich's International Periodicals Directory* (New York: Bowker, 1975–1976 and 1977–1978 editions).

SCIENTIFIC JOURNALS
IN THE UNITED STATES

Chapter 1

Scientific and Technical Information Transfer Through Journals

THE IMPORTANCE OF SCIENTIFIC AND TECHNICAL INFORMATION TRANSFER

Scientific and technical knowledge is one of the most important resources in the world. In recent years, science and technology have made phenomenal advances in electronics, space exploration, medicine, genetics, and numerous other fields. A large portion of these advances were unheard of or thought highly unlikely even a quarter of a century ago. In the early 1950s, it would have been inconceivable to many that humans would have visited the moon, that computers with today's power would be commonplace, or that satellite transmission would make television broadcasts available to remote regions of the globe. Even though scientists and engineers have been able to advance the frontiers of knowledge in many areas, many extremely important problems remain to be solved. In particular, dwindling resources in food, water, air, and energy all present a major challenge to scientific and engineering capability. Many industrial advances have yielded corresponding problems in safety and health. Understanding how to balance and allocate our resources also presents one of our most important challenges.

In the United States, the scientific and technical resource is represented by the knowledge accumulated by 2.8 million scientists and engineers and the estimated 15,000 books, 4,500 journals 4,500 other periodicals,* and thousands of technical reports produced in 1977.†

*By journals, we mean scholarly periodicals that report research and that are often refereed. By other periodicals we mean nonscholarly periodicals such as trade journals,

1

When we add to this the world's living scientists and engineers and its knowledge, current and accumulated in the literature, the knowledge resource is impressive. In the United States alone, we have estimated that 12 million articles have been published since 1839 when the first U.S.-published scientific journal appeared.[2]

The knowledge found in scientists' minds and in the literature is of little use unless it is communicated and assimilated by others. Clearly a great amount of communication must take place for scientists and engineers to accumulate their knowledge. Communication is an essential part of their formal education, as is the continued learning that takes place throughout their professional careers. In their careers, scientists and engineers perform many activities, including research, writing, teaching, self-education, and administration. Information forms an integral part of each of these major activities. Without it scientists and engineers would not be able to function. Research would be almost impossible; writing and teaching would be nonexistent since they are both largely communication functions; formal education and continued learning would also suffer; and development, planning, budgeting, and management would be inadequate. As the vehicle for information, communication is essential to scientists and engineers and to their contribution to society. Thus the various modes of communication become extremely important to science and technology and to society.

Numerous recent studies have demonstrated that scientists and engineers spend an enormous amount of time reading the literature and communicating in other ways. Recently we estimated that U.S. scientists and engineers average about sixty-seven hours per year reading journals. Thus they collectively spent over 200 million hours in 1977 in this activity. Contrary to the belief of some, most scientific and technical journals are thoroughly read. They must have benefit to scientists and engineers, or they would not choose to devote so much of their scarce and valuable time to producing and using them. Their value to society is undoubtedly high. So are the resources expended on scientific and technical communication. We estimate that the total resource expenditures for scientific and technical communication by means of the published literature in the United States are about $12.2 billion. The expenditures related to scientific and technical journals are about $4.7 billion, so it is clear that journals represent a large portion of total resource expenditures of scientific and technical literature.

---

newsletters, and bulletins.

†Machlup refers to these two sources as knowledge in the mind and knowledge in print, respectively.[1]

THE SCIENTIFIC AND TECHNICAL INFORMATION
TRANSFER ENVIRONMENT

Scientists and engineers spend a substantial proportion of their time communicating for a variety of purposes. They use both formal and informal, oral and written means. The modes they use, each suited to particular sets of circumstances, are diverse: face-to-face contact, telephones, teleconferencing, and a host of written materials including letters, memoranda, articles, reports, and books. Evidence suggests that much more communication takes place by the informal than by the formal modes.

Communication may be either direct—represented by items directed to specific individuals such as letters and telephone calls—or indirect. For the most part, oral communication is direct, whether through personal conversations, lectures, or conferences. Indirect channels constitute the bulk of the written literature such as journals, reports, and books. Here potential users of the information contained in the literature must intercept it before they can utilize it. Thus they must go through a two-stage process, first identifying needed literature by such means as scanning, a colleague's reference, or through the secondary literature; then they must actually gain access to the identified document.

The degree of formality of communication also varies. In the informal domain, information is often preliminary and not yet established as reliable scientific knowledge. It is commonly in abstracted form, usually colloquial, frequently incomplete, and sometimes vague. It is often oral and directed at a small audience of associated researchers. Here initial dissemination yields some judgment of the quality of the work, and each recipient uses his or her knowledge of the field to evaluate the work. As the work is presented and evaluation takes place, manuscripts are shaped and reshaped, and the type of reporting used becomes increasingly formalized. After this process, formal publication is made, usually in the form of a journal article, now presented to a more universal audience as a finished scientific product.

One approach to describing many of the communication activities of scientists and engineers is to trace the various reportings of a particular research project. Such reports generally begin while the research is in progress and may continue for several years after the research has been completed. The evolution of these reportings, and in particular their timing, has been studied in some detail for a number of disciplines by the Center for Research in Scientific Communication of the Johns Hopkins University.[3]

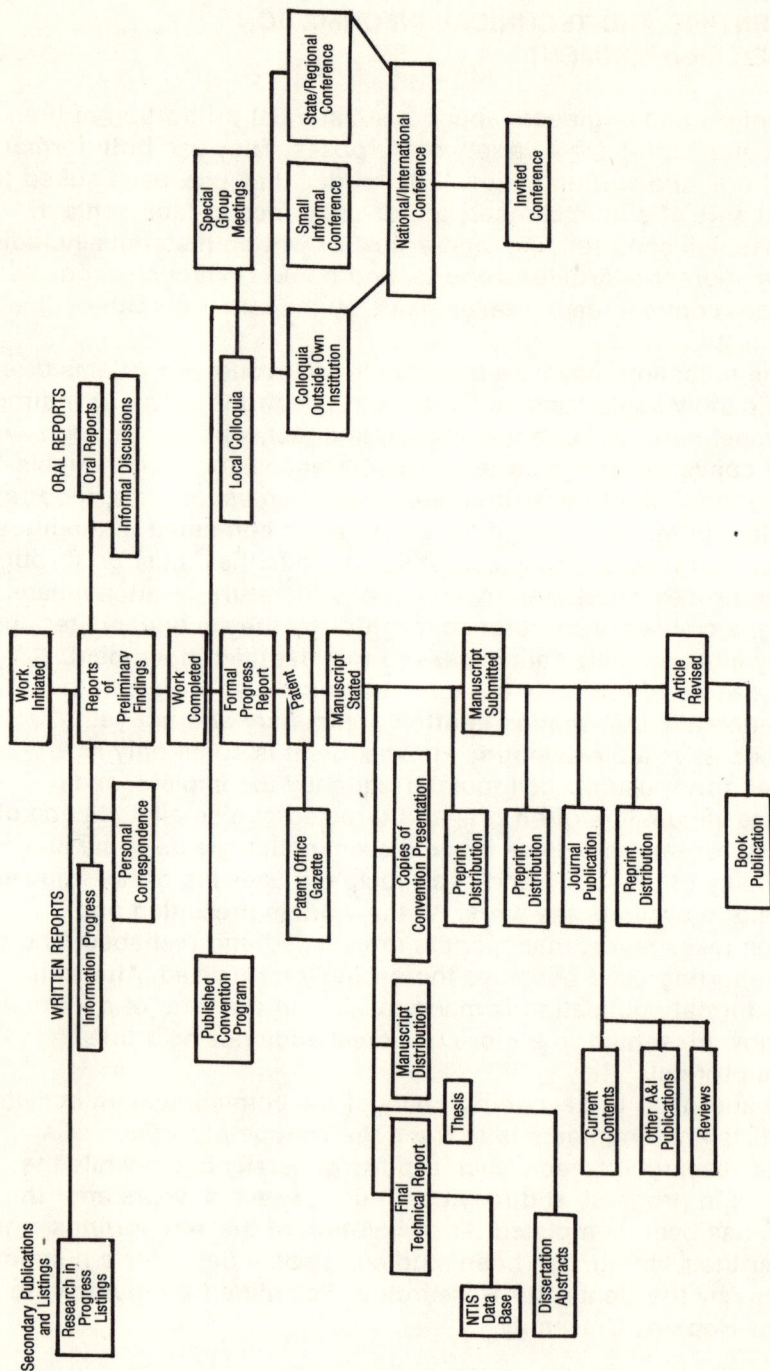

Figure 1.1 The dissemination of scientific and technical research results. Adapted from W. D. Garvey et al., *An Overview of the Information Exchange Associated with National and Scientific Meetings in Relation to the General Process of Communications in Science* (Baltimore: Center for Research in Scientific Communication, Johns Hopkins University, 1970).

4

Figure 1.1 illustrates the major components of the research report cycle.* It begins with the initiation of a particular project and continues through the presentation of both oral and written reports and through the creation of secondary publications and listings. The average elapsed time from the initiation of work to the publication of a journal article is about two to two and one-half years, depending on the field. Publication of a book reporting the same results may lag behind publication of a journal article by several years. Although timing is not indicated precisely in Figure 1.1 and not all modes of communication occur in any one instance of reporting of a research finding, the sequence of events does represent that generally found in scientific and technical communication.

Some reports precede the completion of work. These may be oral or written but tend toward informal progress reports, presentations to colleagues, and the like. Twenty percent of the authors (of journal articles) studied by the center had reported preliminary findings in some form.

The majority of prepublication dissemination takes place after work is completed and is ready to be reported. In fact, some 80 percent of authors make one or more prepublication reports, with two-thirds making oral and one-half making written reports. Almost half of the authors make both oral and written reports. For all authors, the average number of prepublication reports was 2.4. These figures vary somewhat according to the scientist's field.

The type of oral prepublication report presented ranges from informal presentations, to small, in-house groups, to formal reports given at national or international meetings. Written prepublication reports include personal correspondence, technical reports, theses, in-house publications, patents, proceedings and copies of an oral presentation, and other journal articles.† In general, there is a sequential organization of events starting with the more highly focused audiences and ending with the more general ones. This sequence seems especially suited to help authors shape their work gradually and effectively for journal publication. In fact, about half of the authors who made prepublication reports indicate that they received feedback because of those reports, and that this information led them to modify the presentation of their work in the journal manuscripts they submitted later.

*In some cases the author of a particular written report need not be the performing scientist, as in institutional progress reports, trade journal news items, or monographs about recent advances in the field.
†Some results may be reported only in prepublication reports.

Prepublication reports are generally presented or written until the time a journal article manuscript is submitted for publication. This means that prepublication reporting takes place over about a seven-month period, starting fifteen months before publication.* In the remaining eight months before publication, the primary form of reporting is dissemination of article preprints.† These are distributed both before and after article submission and before and after the article is accepted for publication. Somewhat over half the authors of articles distribute preprints at least at one stage; for these scientists, the average number of preprints is slightly less than nine. Like other prepublication reports, preprints are often used as a method of obtaining feedback, which leads to modifications of the journal article.

The first public distribution of research results comes when an article is finally published, some two to two and one-half years after the research was initiated. This period may be somewhat longer if the article is rejected by the first publisher to which it is submitted, and one or more additional submissions are made. By the time of publication, the scientists as authors are likely to have already begun preliminary reporting on their next research venture and possibly to have begun a third piece of research. Thus the time lag between completion of work and formal publication is about the same as a full research cycle.

After initial journal publication, the findings may be reported in other journal articles, perhaps as many as eight.[4] Distribution of all the articles takes place over a period of time through personal subscriptions, library copies, or author reprints. Following journal publication, a portion of the research findings are ultimately reported in scientific and technical books. This often happens anywhere from four to thirteen years following the beginning of research.

An additional factor to be noted in Figure 1.1 is the production of secondary publications and listings. These are developed in a variety of formats, and they call attention to the research results at virtually every stage and mode of reporting. Thus they provide a valuable means of access, both prior to and after journal publication, to the literature. In this way, they extend early access beyond a core group to a more general public.

Overall the reporting cycle reflects a variety of modes of presentation, including direct and indirect, formal and informal, and oral and written. As the cycle progresses, the size of the audience reached and the degree of formality increase. With publication of a journal

*Patents are an exception to this time frame. Patient applications are normally filed before journal submission, but the patent itself may not be issued until several years later.
†We define a preprint as any copy of an article manuscript (before or after acceptance by a journal) that is distributed prior to publication and distribution in a journal.

article, the research results are certified and made available to the entire scientific community. There is redundancy in this cycle, but for the most part it appears to be a necessary part of the interactive social process leading to general acceptance of scientific research results.

Journal publishing represents the heart of the scientific and technical communication. It is the most extensive mode found in the published literature and represents the greatest amount of resources. For these reasons, we devote most of the remainder of this book to scientific and technical journals published in the United States.

THE HISTORY OF SCIENTIFIC AND TECHNICAL JOURNALS

Scientific and technical journals are now over three centuries old. According to Houghton,[5] the first scientific journal, the *Journal des Scavans,* was founded in 1665 by de Sallo, a counsellor of the French court of the Parliament. This first issue, twenty pages long, contained ten articles, some letters, and notes. In the same year an eminent group of English philosophers published an English scientific journal, *Philosophical Transactions.* Scientific journals were also published in Germany and other parts of Europe soon after. The first scientific journal was published in the United States in 1839.[6] The growth of scientific and technical journal publication worldwide and in the United States has been remarkable since the beginning of publishing.*

It is estimated that by 1800 there were about ninety scientific journals worldwide, and this number increased to over ten thousand by 1900. Price has plotted the growth of number of journals since their inception (Figure 1.2).[10] He points out that the number of journals doubles about every fifteen years and that this rate of growth has been relatively constant over the centuries. This rate represents an average growth of about 5 percent a year. It is found to be nearly linear on a logarithmic scale.

As the number of scientific journals began to proliferate worldwide, it became necessary to develop a means of identifying and locating articles. This requirement led to the development of abstract journals, which were first published in the mid-1800s. There were about three

---

*Both Houghton[7] and Meadows[8] have recently described the character of this growth through its first two centuries. Earlier Kronick[9] provided a historical account of scientific journals.

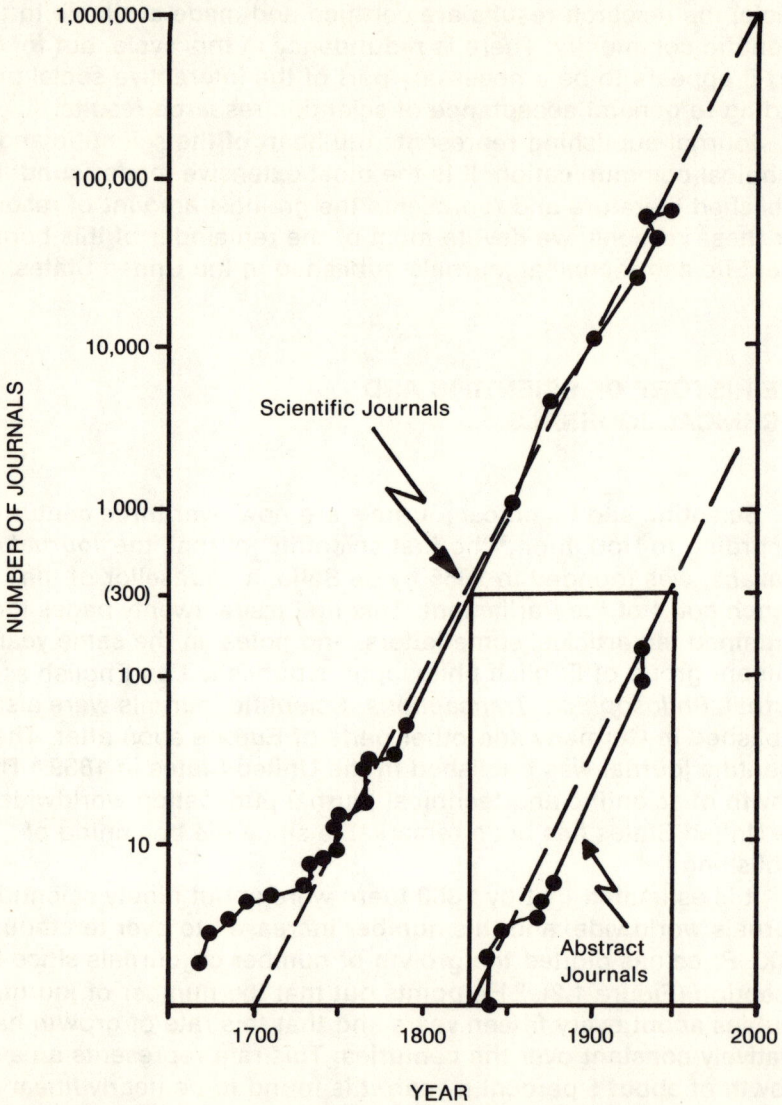

Figure 1.2 Total number of scientific journals and abstract journals founded worldwide, 1665–2000. From Derek de Solla Price, *Little Science Big Science* (New York: Columbia University Press, 1979) p. 9.

hundred scientific journals worldwide at that time, and there has been about one new abstract journal for every three hundred new scientific journals. The parallel growth (in a logarithmic scale) of scientific

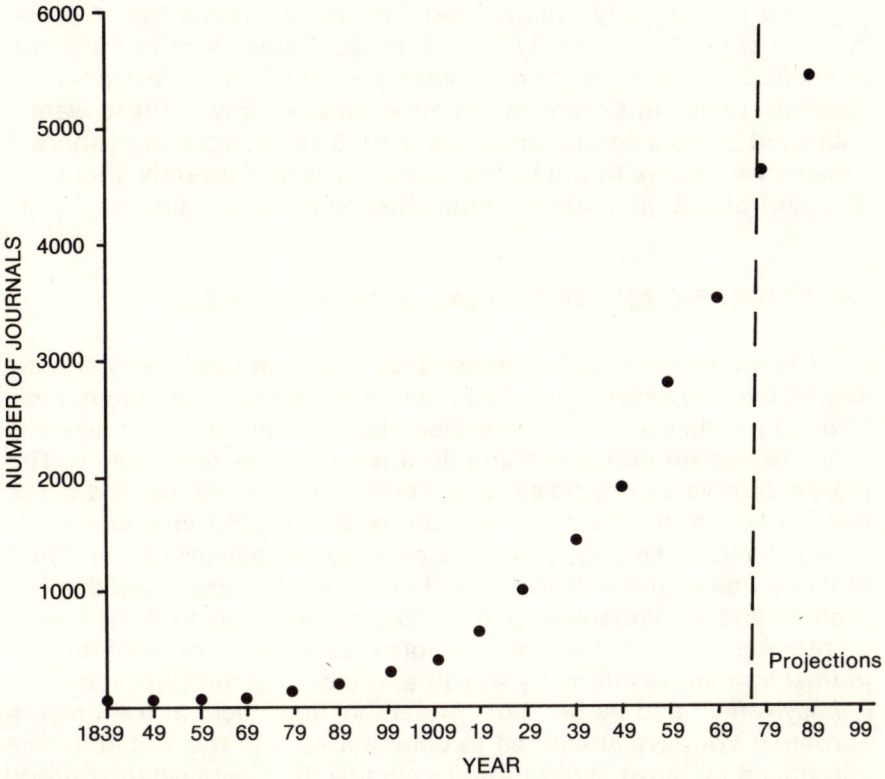

Figure 1.3 Total number of scientific and technical journals in the United States, 1839–2000.

journals and abstract journals as demonstrated by Price is shown in Figure 1.2.

The growth of journals in the United States has also been very rapid, but in recent years it has been less than worldwide growth. The growth of the number of journals in the United States, displayed in Figure 1.3, also has had a doubling every fifteen years until recently. The growth was more rapid than that observed worldwide up to about 1875 and then parallel to the worldwide growth until the Great Depression in the 1930s. From that time, the rate appears to be steady and linear on the logarithmic scale.

Two other factors are part of the picture of scientific and technical publishing in the United States. The first is the entry of scientific and technical periodicals that do not report primary research findings. They include trade journals, bulletins, newsletters, and the like. In 1977 it is estimated that there were about an equal number of scientific and technical journals and these other types of periodicals. Their number is slightly less than forty-five hundred. The second factor has to do with the type of publishing entity. In the United States, prior to 1945 nearly all scientific and technical journals were published by professional societies, unlike in European countries where many of these were published by commercial firms. After 1945 commercial publishers became very active in the United States as well. Currently about one-third of U.S. journals are published by commercial firms.

THE SCIENTIFIC AND TECHNICAL JOURNAL SYSTEM

The literature is full of studies that report on publishers, libraries, or journal use. However, there had been little research on communication through journals as an entire system, but recently some studies have concentrated on publishers and libraries and their relationship. The most extensive such studies were performed by Machlup and Leeson[11] and Fry and White,[12] and at the National Enquiry.[13] Two major studies have examined the possibility of a national periodicals center. The first of these was by the National Task Force on a National Periodicals System, and a companion report was prepared by the Council on Library Resources.[14,15] A more complete description of the entire journal system was done by Ackoff and others at the University of Pennsylvania[16] and by Senders, Anderson, and Hecht at the University of Toronto.[17] We have attempted to combine each of the system elements introduced by these studies and to employ their data when appropriate. However, we have also broadened the journal system concept to include those system functions and activities associated with information generated through research and development, authorship, publishing, librarianship, data-base generation and distribution, on-line searching, and use of scientific and technical information.

A highly significant study was performed under a National Science Foundation grant by Machlup and Leeson.[18] Machlup and Leeson were concerned with the economics of scientific and technical information distributed through journal publications, primarily scholarly books and journals. They performed an in-depth case study of publishers that were primarily located in the northeastern part of the United States. This study is clearly the most exhaustive one performed on the financial, economic, and cost aspects of scientific and technical publishing. We have relied heavily on this part of their study to provide source data and to validate some of our results. Much of these data were gathered by personal interview and inspection of publisher records. The researchers also obtained a considerable amount of financial, economic, and cost data on libraries, wholesalers, and abstracting and indexing services. A study involving use of economic journals was also performed by the study team. Some results of this extensive effort have been incorporated into their publication.

Another series of studies on libraries and publishers conducted by Fry and White at Indiana University report a series of percent changes in volume and type of library operations relating to scholarly publications. The trends include observations from 1969 in two-year intervals to 1973 and then annual observations to 1976. The researchers did a survey of publishers but suffered from a very high nonresponse, which makes their results questionable, although they do provide some useful evidence concerning publishing activities. (Every research endeavor that we are aware of has encountered this problem of lack of cooperation by publishers.) We have used these data extensively to help verify some of our results and observed trends. Another study was recently completed on scholarly knowledge. This study, which relies heavily on the data sources above, as well as our work, provides a number of recommendations for the humanities.

A national periodicals system was proposed by the Task Force of the National Commission on Libraries and Information Science. The Council on Library Resources prepared a technical development plan for a center that would be the heart of this system. The system consists of a loosely defined hierarchical network of libraries. There would be a National Periodicals Center that accumulates titles currently published and back-files of current titles. Requests for no-longer-published titles would be obtained through agreements with other organizations. Only about 20 percent of the requests would be satisfied by the center; the balance would be handled by individual libraries or through state or regional systems.

The SCATT Report provides a description of an idealized national scientific communication and technology transfer (SCATT) system. This system is designed to facilitate the flow of scientific and technical information found (1) in several forms such as audio (formal such as a prepared lecture or informal such as a personal conversation) and visual (formal such as reading a published article or informal such as an exchange of personal letters); (2) at three message levels: primary (messages believed to convey new information), secondary (messages about primary messages), and tertiary (messages about the content of other messages); and (3) by four functions involving recorded messages (the production, dissemination, acquisition, and use of the messages). The project attempted "to mobilize the large number of relatively autonomous subsystems of the current system into a collaborative effort directed at redesigning their system and implementing their design. The

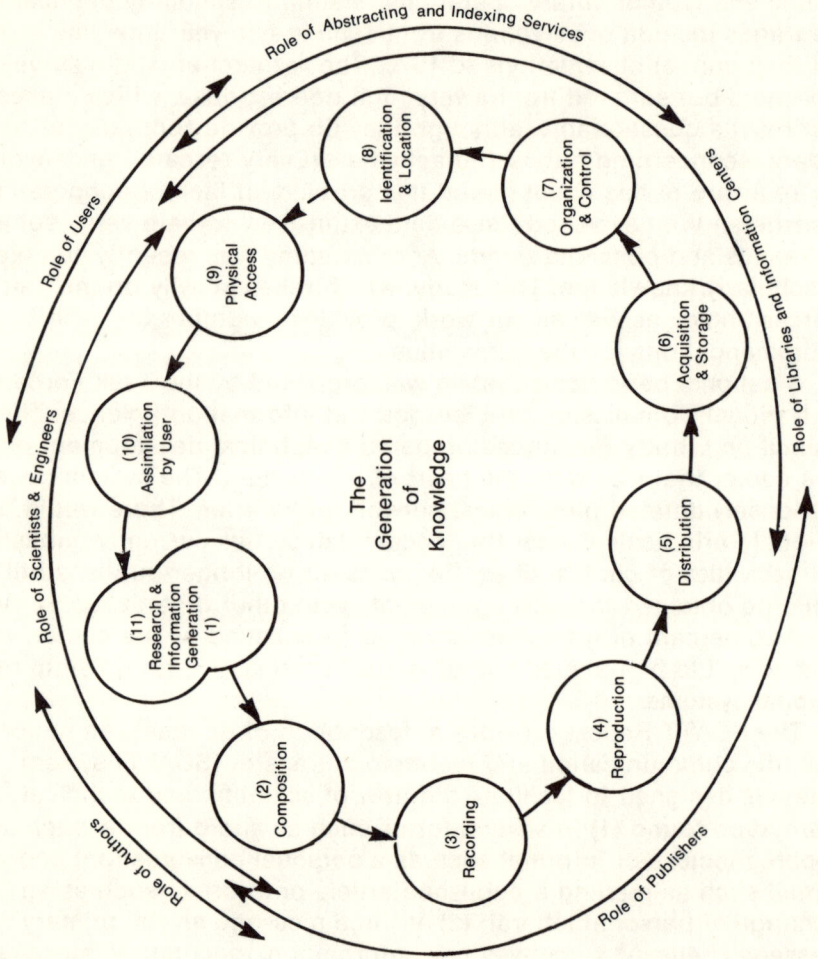

Figure 1.4 Scientific and technical information transfer model.

(1) Research & Information Generation
(2) Composition
(3) Recording
(4) Reproduction
(5) Distribution
(6) Acquisition & Storage
(7) Organization & Control
(8) Identification & Location
(9) Physical Access
(10) Assimilation by User
(11) Research & Information Generation

The Generation of Knowledge

Role of Abstracting and Indexing Services
Role of Users
Role of Scientists & Engineers
Role of Authors
Role of Publishers
Role of Libraries and Information Centers

idealized design process used is intended to stimulate such a mobilization, and the design produced is intended to serve as a platform from which that mobilization can spring."

Senders and colleagues, concentrating primarily on journal publishing, reported on a scientific publication system. Their background work included application of an industrial dynamics model to the journal publication system, which permits one to determine the interactive effects of several participants in the system. The researchers looked at past trends and the future of journal systems with an emphasis on electronic processes and other future technology. They feel that an electronic alternative to paper-based journal publishing is inevitable.

All of these studies emphasize the importance of scientific and technical journals in communication. For this reason, we too have concentrated on describing the journal system. Since their inception three centuries ago, journals have been the major scientific and technical communication vehicle and have fulfilled three basic roles: the archival role, the dissemination or communication role, and the social role of conveying prestige and recognition.[19] In order for the journals to fulfill these roles, several functions must be performed by a broad range of participants. Because of the diversity of these functions and the participants involved, communication by journals is not easily characterized. The journal system consists of many facets, including numerous types of information found in many forms, involving many activities, through a large number of processes performed by a wide spectrum of participants through a myriad of communication paths. The system structure is very complex, often governed by conflicting laws and government regulations.* The motives and incentives of the many different participants conflict. The interfaces among them are subtle and therefore important to understand fully. To complicate this picture further, the journal system resides in an environment of constantly changing technology.

Some of the purposes of this book are to sort out this complex system, to facilitate an understanding of it, and to apply this understanding to improve its operation and system structure. We have formulated a quantitative model that describes this system in terms of its cost and flow of materials. The heart of the model is a schema of communication of science and technology (Figure 1.4).† This diagram

---

*For example, government-sponsored studies have indicated the contribution made by page charges to the support of the journal system.[20] Yet the Postal Service[21] and the Internal Revenue Service have both suggested rulings that would hamper that system.
†This schema was first described in *Statistical Indicators of Scientific and Technical Communication: 1960–1980.*[22]

represents an information transfer spiral based on published documents, although some of the functions described in it are applicable to other forms of communication as well, such as through conferences or the communication of bibliographic information through publications or computer searches. The spiral includes ten functions that are essential to complete the transfer of information. These functions were chosen because they appear to be discrete in their representation of the communication processes and in the statistical and economic factors related to these processes.

It is convenient to consider the spiral as beginning at research, which results in generation of knowledge (1). This function is the role of scientists and engineers. Obviously without research results to report, there would be little need for communications media. The quantity of new research results and the amount of subsequent use of them are dependent on the number of scientists and engineers and on the resources expended on research.

As a result of scientific research projects, manuscripts (books, articles, reports) are composed (2). The composition function refers to formal writing, editing, and reviewing of the manuscripts. When a manuscript is in a form to be communicated, it is recorded (3). These two functions are the role of authors, publishers, editors, and reviewers. At this stage authors have as yet very little impact on the scientific community by means of formal communication. Only when the work has been reproduced and distributed does it gain the potential for widespread influence on an audience.

The reproduction (4) and distribution (5) functions are usually the role of the publishers, however, authors, libraries, and colleagues also play an important role. The transfer of documents through the three participants may be thought of as indirect reproduction and distribution, which requires acquisition and storage (6). Although many individuals acquire scientific and technical books, articles, or reports and may store them, this stage of the spiral is represented by libraries and other information centers. Through their acquisitions and storage policies, libraries provide a permanent archive of scientific achievement.* They also ensure access to this record.

Libraries also have an important role to play in organization and control (7) functions. In addition to collecting publications, libraries and other information centers provide access to these documents through cataloging, classifying, indexing, and other related procedures. The major indexing and abstracting services and bibliographic services play

*The archive is not restricted to books and journals; it often includes theses, technical reports, patents, convention programs, data bases, and bibliographic publications.

an important part in organization and control as well. Needed publications may be identified and located (8) by a number of processes, including reference to one's own subscription, library search, and computerized search and retrieval systems. This function is often accomplished for users by an intermediary from a library or other information service. The physical access (9) function includes direct distribution of scientific and technical articles from publishers to users, indirect distribution through libraries and other information centers, and distribution of reprints by the author. The final function in the spiral, assimilation by User (10), is the least tangible. It is the stage at which information (as opposed to documents) is transferred. It is at this stage that the state of the user's knowledge is altered.

The functional schema is presented as a spiral because the communications process is continuous and regenerative. Through assimilation readers may gain information that can be used in their research in such phases as conceptualization, design, experimentation, and analysis. This research may, in turn, generate new composition and recording for another cycle through the information transfer spiral.

The notion of describing information systems by functions is not at all new. Other researchers such as Berul,[23] Doyle,[24] Hayes and Becker,[25] Kent,[26] King and Bryant,[27] Krevitt,[28] Simpson and Flanagan,[29] Stevens,[30] Wall,[31] and Weisman[32] have also provided a partial list of functions that are related to the ones mentioned above. These functions are discussed in greater detail by King et al.[33] They are given in tabular form in appendix A of that report.

A further function, which is implicit in the communication schema, is that of transmission. An examination of the flow involved in the system of functions in Figure 1.4 makes it evident that there are many transmissions of information among participants. One can aggregate these transmissions into those that provide flow between participants performing two generic functions such as composition and recording. Most transmissions in the composition and recording functions include mailing of manuscripts to publishers and transmission of letters and manuscripts back to authors depending on the status of the submitted manuscript. A complete description of transmissions among composition and recording participants is very complex.

Once a manuscript is accepted, the publisher makes a master record and replicates copies in the form of journal issues, reprints, or microform for distribution to users, libraries, and other entities such as abstracting and indexing services (distribution function). The user may read an article upon receipt but normally will store the information for later use. Thus most transmissions of copies from participants in the recording function (publishers) are to participants in the acquisition and

16

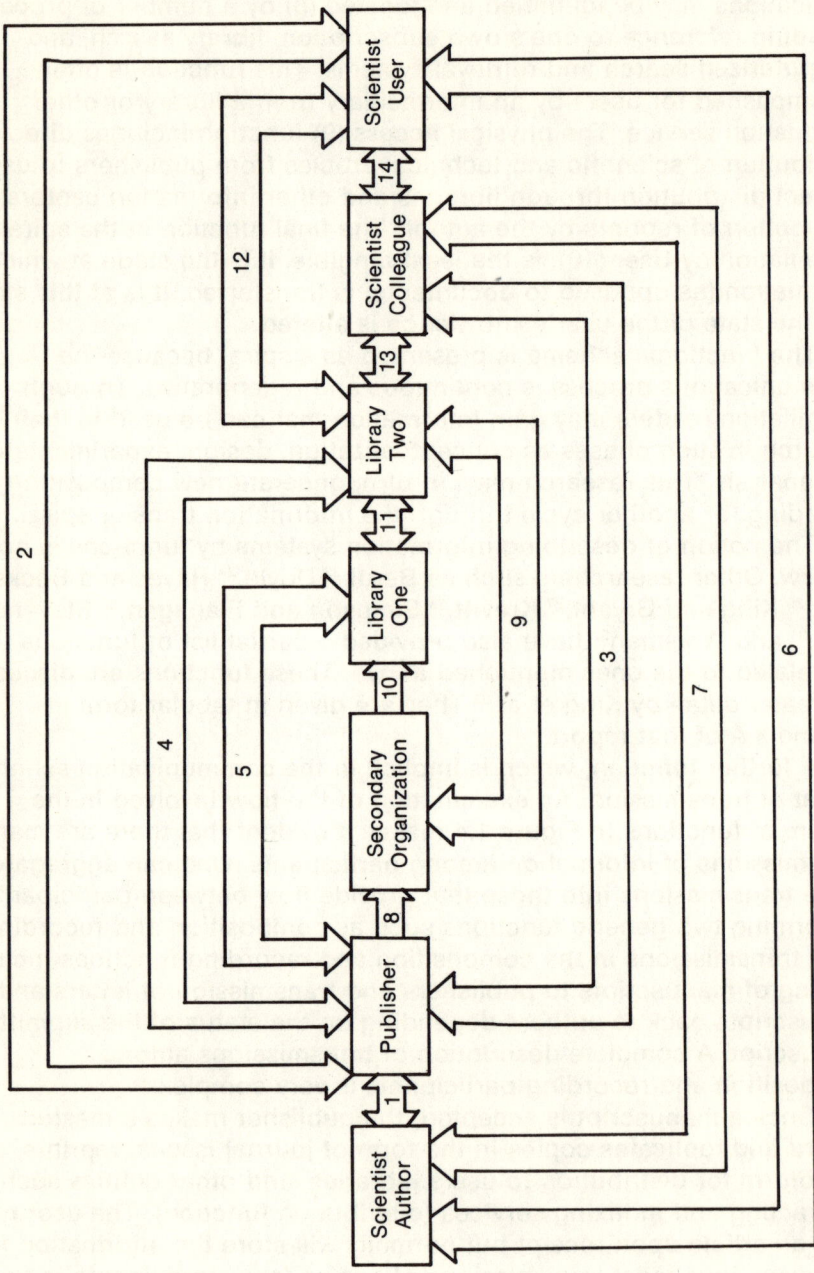

Figure 1.5 General flow of information among scientists and engineers.

storage function (scientists, libraries, and abstracting and indexing services). Users can obtain information from any one of a number of participants in the acquisition and storage function, including themselves, libraries, and authors. The information can be in several forms, such as journal issues, photocopies, reprints, or microform. Some transmissions are library requests to publishers for copies or subscriptions and reprint requests from users to authors.

When one begins to link all the transmissions for specific messages as they flow from authors to users, several paths together emerge as dominant in the number of transmissions and ultimate uses. One dominant path is from authors to publishers and then to users. Another is from authors to publishers to libraries and then to users. The third is from the author to publishers to authors (reprints) and then to users. The major paths are displayed in Figure 1.5, with all of the principal participants shown as nodes in the flow (scientists as authors, colleagues, and users; publishers; libraries; secondary organizations).

The flow of information from one scientist to another can take a number of paths depending on the basic uses made of the information and the form in which it is transferred. For example, communication can be direct from one scientist (author) to another (user) through path 6 or by a colleague through paths 7 and 14 in the form of informal communication, conferences, or exchange of unpublished technical reports as well as preprints and reprints of published articles. The flow can be through formal publishing in scholarly journals and books, in which case manuscripts will go from authors to publishers (path 1), and the publishers in turn will distribute articles and books directly to scientists (paths 2 and 3) or through libraries (paths 4 and 5), which in turn store and provide the information to scientists (paths 12 and 13). Publishers also distribute copies to secondary organizations (path 8) that abstract and index journal articles. Their publications are distributed mostly to libraries (path 10) that use them for searching, organization, and control. In increasing numbers, libraries rely on other libraries to provide copies through interlibrary loan (path 11). Here we estimate the extent of flow of materials through each of the fourteen paths mentioned above, as well as through other transmissions.

So far we have shown that there are several generic functions that must be performed in a journal system (Figure 1.4); that there are four principal participants (scientists as authors and readers; publishers; libraries; and secondary organizations), and that flow of information can be aggregated into a reasonable number of paths and counts made of materials involved and number of uses occurring through each path. Another useful categorization is that of specific information products and services that are involved in the system, such as journals, interlibrary loans, and on-line search systems. These many products and services form the basis for most of the actual information messages* that are communicated in the system. These messages may be fully descriptive information such as that found in journal articles, books, or technical reports, or they may be messages about the fully descriptive information such as those found in bibliographic data bases, abstracting and indexing publications, or citations of primary publications.

In order to prepare an information product or to provide a service, many specific activities must be performed. In publishing a journal, for example, activities include technical editing, copy editing, graphics preparation, proofreading, typesetting, engraving, printing, collating, binding, wrapping, and mailing. Each activity can be described by the processes employed, as well as by the units of input and output. All information products and services are the result of many such activities.

Thus far we have merely described the journal system by a categorization that subdivides the system into discrete and definable components. These components, however, form the basis for further modeling the system in terms of its cost. The basis for establishing the journal system cost is the cost of specific activities. From these, the costs of products and services, functions, and finally the entire system are built up. By summing the cost of hundreds of activities, we arrive at a gross estimate of the cost of the entire journal system. Since we also have arrived at estimates of the number of readings of journal articles, we can estimate the cost per reading associated with the journal system in its entirety or through segments of it. The framework of the journal system that we describe above serves as the basis for the analysis presented in this book.

REFERENCES

1. F. Machlup and K. W. Leeson, *Information Through the Printed Word: The Dissemination of Scholarly, Scientific and Intellectual Knowledge,* vol. 1: *Book Publishing;* vol. 2: *Journals;* vol. 3: *Libraries* (New York: Praeger, 1978).
2. National Science Foundation, *Characteristics of Scientific Journals, 1949-1959* (Washington, D.C.: U.S. Government Printing Office, 1964).
3. W. D. Garvey et al., *An Overview of the Information Exchange Associated with National Scientific Meetings in Relation to the*

*As distinct from functional messages such as reprint requests or acceptance notices.

General Process of Communication in Science (Baltimore: Center for Research in Scientific Communication, Johns Hopkins University, 1970).

4. W. D. Garvey and S. D. Gottfredson, "Scientific Communication as an Interactive Social Process." Paper presented to the U.S./U.S.S.R. Symposium, Yale University, October 1975.

5. B. Houghton, *Scientific Periodicals* (Hamden, Connecticut: Linnet Books, 1975).

6. National Science Foundation, *Characteristics of Scientific Journals.*

7. Houghton, *Scientific Periodicals.*

8. A. J. Meadows, *Communication in Science* (London: Butterworths, 1974).

9. D. A. Kronick, *A History of Scientific and Technical Periodicals* (New York: Scarecrow Press, 1962).

10. D. S. Price, *Little Science Big Science* (New York: Columbia University Press, 1963).

11. Machlup and Leeson, *Information Through the Printed Word* vol. 2.

12. B. M. Fry and H. S. White, *Publishers and Libraries: A Study of Scholarly and Research Journals* (Lexington, Mass.: Lexington Books, 1976).

13. *Scholarly Communication: The Report of the National Enquiry* (Baltimore: Johns Hopkins University Press, 1979).

14. Task Force on a National Periodicals System, *Effective Access to the Periodical Literature: A National Program* (Washington, D.C.: National Commission on Libraries and Information Science, 1977).

15. Council on Library Resources, *A National Periodicals Center: Technical Development Plan* (Washington, D.C.: Council on Library Resources, 1978).

16. R. L. Ackoff, T. A. Cowan, W. M. Sachs, M. L. Meditz, P. Davis, J. C. Emery, and M. C. J. Elton, *The SCATT Report: Designing a National Scientific and Technological Communication System* (Philadelphia: University of Pennsylvania Press, 1976).

17. J. W. Senders, C. M. B. Anderson, and C. P. Hecht, *Scientific Publication Systems: An Analysis of Past, Present, and Future Methods of Scientific Communication* (Toronto: University of Toronto, 1975).

18. Machlup and Leeson, *Information Through the Printed Word.*

19. A. Herschman, "The Primary Journal: Past, Present, and Future" (Paper presented at Division of Chemical Literature Symposium on Primary Journals, 157th Annual Meeting of American Chemical Society, Minneapolis, Minnesota, April 1969).

20. Capital Systems Group, *Page-Charge Policies and Practices in Scientific and Technical Publishing: A Historical Summary and*

Annotated Bibliography (Rockville, Maryland: Capital Systems Group, 1976).

21. 1972 U.S. Postal Service interpretation of PL 86-682 (Washington, D.C.: 86th Congress, September 2, 1960).

22. D. W. King, D. D. McDonald, N. K. Roderer, and B. L. Wood, *Statistical Indicators of Scientific and Technical Communication (1960–1980), vol. 1: A Summary Report* (Washington, D.C.: Government Printing Office, 1976).

23. L. Berul, *Information Storage and Retrieval—A State of the Art Report* (Philadelphia: Auerbach Corp., 1964).

24. L. B. Doyle, *Information Retrieval and Processing* (Los Angeles: Melville Publishing, 1975).

25. R. M. Hayes and J. Becker, *Handbook of Data Processing for Libraries* (New York: Wiley, 1970).

26. A. Kent, *Specialized Information Centers* (Washington D.C.: Spartan Books, 1965).

27. D. W. King and E. C. Bryant, *The Evaluation of Information Services and Products* (Washington, D.C.: Information Resources Press, 1971).

28. B. I. Krevitt, *Editorial Processing Centers: A Study to Determine Economic and Technical Feasibility* Annex Part I: A Baseline Study of Current Journal Publishing Practices in the Life Sciences (Rockville, Maryland: Westat, and Aspen Systems Corporation, 1974).

29. G. S. Simpson, Jr., and C. Flanagan, "Information Centers and Services," in *Annual Review of Information Science and Technology,* ed. Carlos A. Caudra (New York: Wiley-Interscience, 1966).

30. M. E. Stevens, *Research and Development in the Computer and Information Sciences,* vol. 2: *Processing, Storage, and Output Requirements in Information Processing Systems: A Selective Literature Review,* NBS Monograph No. 113 (Washington, D.C.: National Bureau of Standards, 1970).

31. E. Wall, "A Rationale for Attacking Information Problems," *American Documentation* 18:97–103 (1967).

32. H. M. Weisman, *Information Systems, Services and Centers* (New York: Wiley, 1972).

33. D. W. King, N. K. Roderer, C. K. Mick, and R. H. Miller, *Systems Analysis of Scientific and Technical Communication in the United States: Annex 1. Communication Functions in Science and Technology* (Rockville, Maryland: King Research, 1978).

Chapter 2

Summary Analysis

THE SCIENTIFIC AND TECHNICAL COMMUNICATION ENVIRONMENT

There are many reasons for the scientific and technical community's major breakthroughs in recent years. Certainly one of the most important is the achievement of knowledge through scientific and technical communication, an integral part of scientists' and engineers' research activities. Recent studies demonstrate that scientists and engineers on the average spend between 13 and 17 percent of their working time writing, between 8 and 14 percent of their time reading, and between 8 and 26 percent of their time in oral communication.

The communication processes involving scientific and technical discovery of research results take many forms and occur at many levels of formality. Garvey and his associates[1,2,3] have described these processes in great detail. Such communication begins very informally with personal discussions during research and progresses to a more formal oral or written report of preliminary findings. After the research is completed, the results are often presented at local or special group meetings and then at major professional conventions. Most often, at this juncture the results are prepared for formal publication in technical reports and journals. Later they may appear in books.

The heart of this overall process seems to be the journal literature. It is the first place where the new findings are presented in a permanent record that is widely distributed, identified through extensive bibliographic products and services, and made available through libraries and other information services.

THE GROWTH OF THE SCIENTIFIC AND TECHNICAL LITERATURE

The amount of scientific and technical literature has been increasing steadily since 1960 and is expected to continue to do so through 1985. The numbers of periodicals, journals, books, and journal articles published since 1960 are given in Figure 2.1. Much of the growth

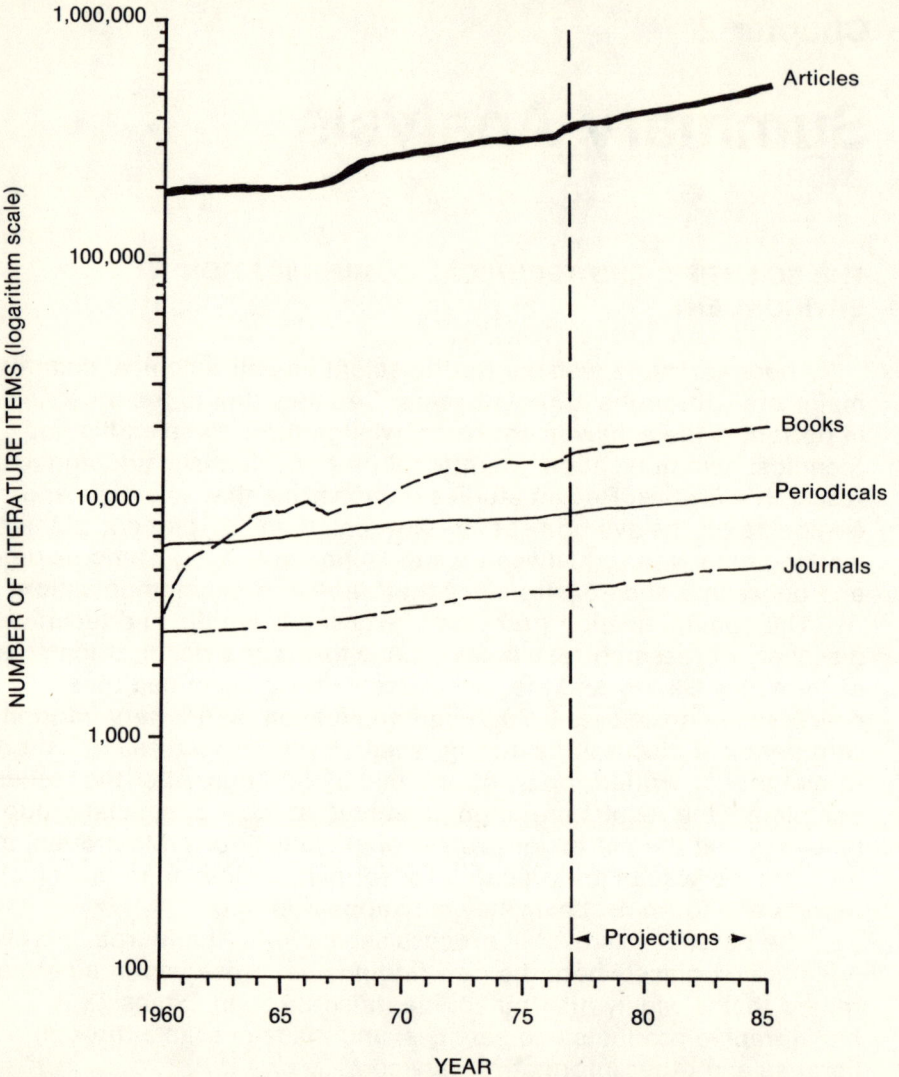

Figure 2.1 Number of literature items published, 1960–1985.

in U.S. scientific and technical communications is due to growth in the number of scientists and engineers (Figure 2.2). There were an estimated 2.8 million scientists and engineers in 1977, up from 1.9 million in 1960. All of the literature forms show steady growth, but a different picture emerges for the average number of items per scientist or engineer (Figure 2.3). The number of scientific and technical books published per one thousand scientists increased sharply in the early 1960s and then leveled off. Periodicals other than the journals have

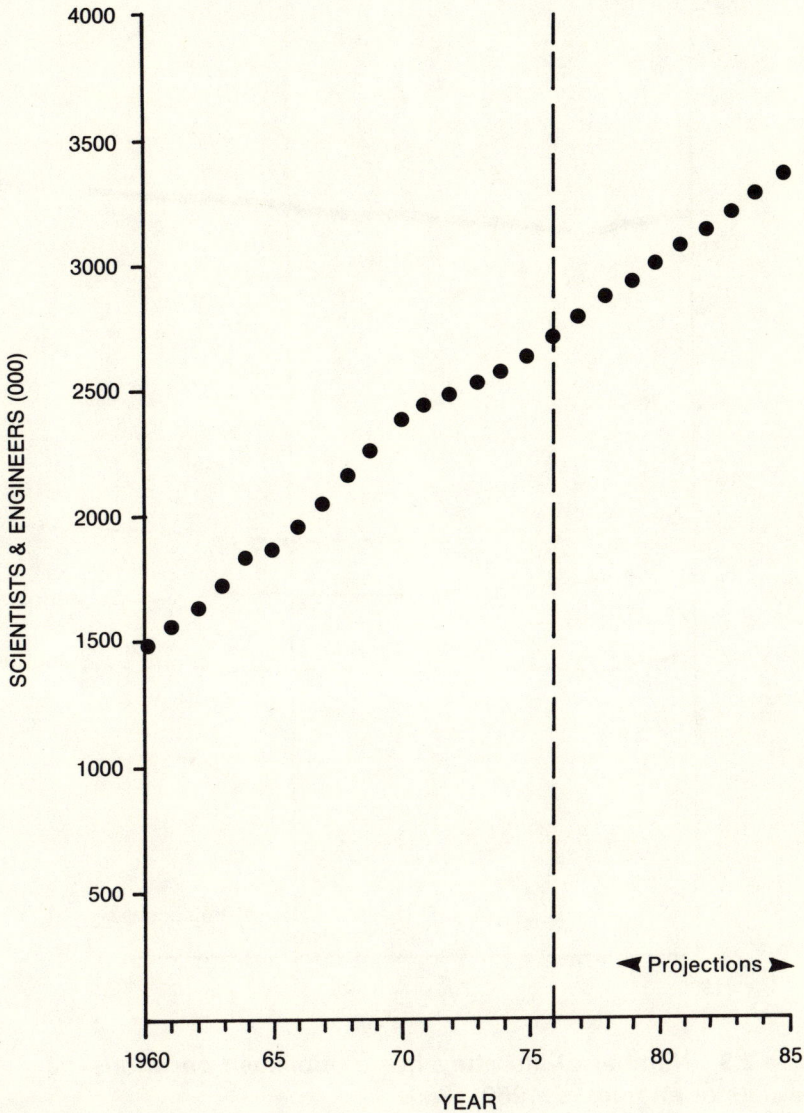

Figure 2.2 Number of scientists and engineers over time, 1960–1985.

decreased somewhat. Journals, on the other hand, have maintained an almost constant rate over this period.

The number of periodicals published worldwide is estimated to be 57,400 compared to the 8,915 periodicals and 4,447 journals published in the United States in 1977. The trends of literature publishing

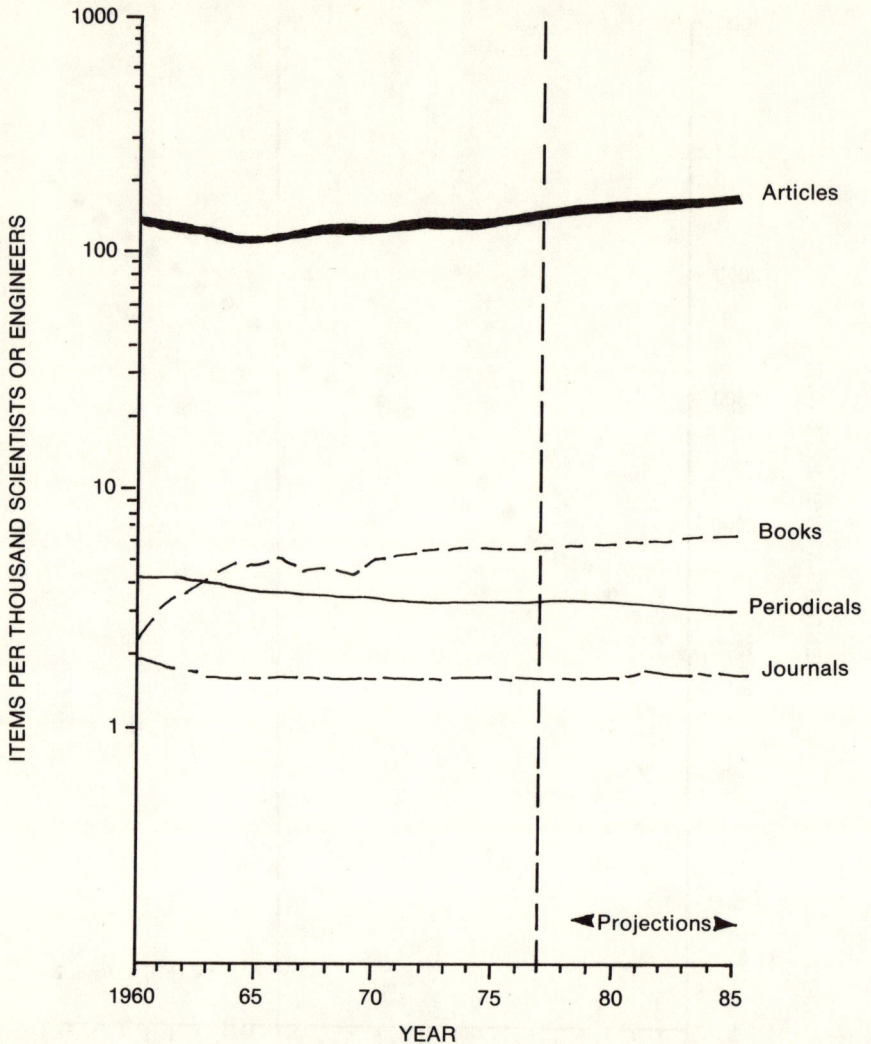

Figure 2.3 Number of literature items published per thousand scientists or engineers, 1960–1985.

worldwide and in the United States are shown in Figure 2.4. The rate of growth in the number of scientific and technical periodicals and journals has been less in the United States than worldwide. Price[4] has demonstrated that the growth of science and scientific literature is highly correlated with indicators of economic growth of nations. Since many developing countries are currently in rapid growth, the journal literature appears to be correlated to this growth.

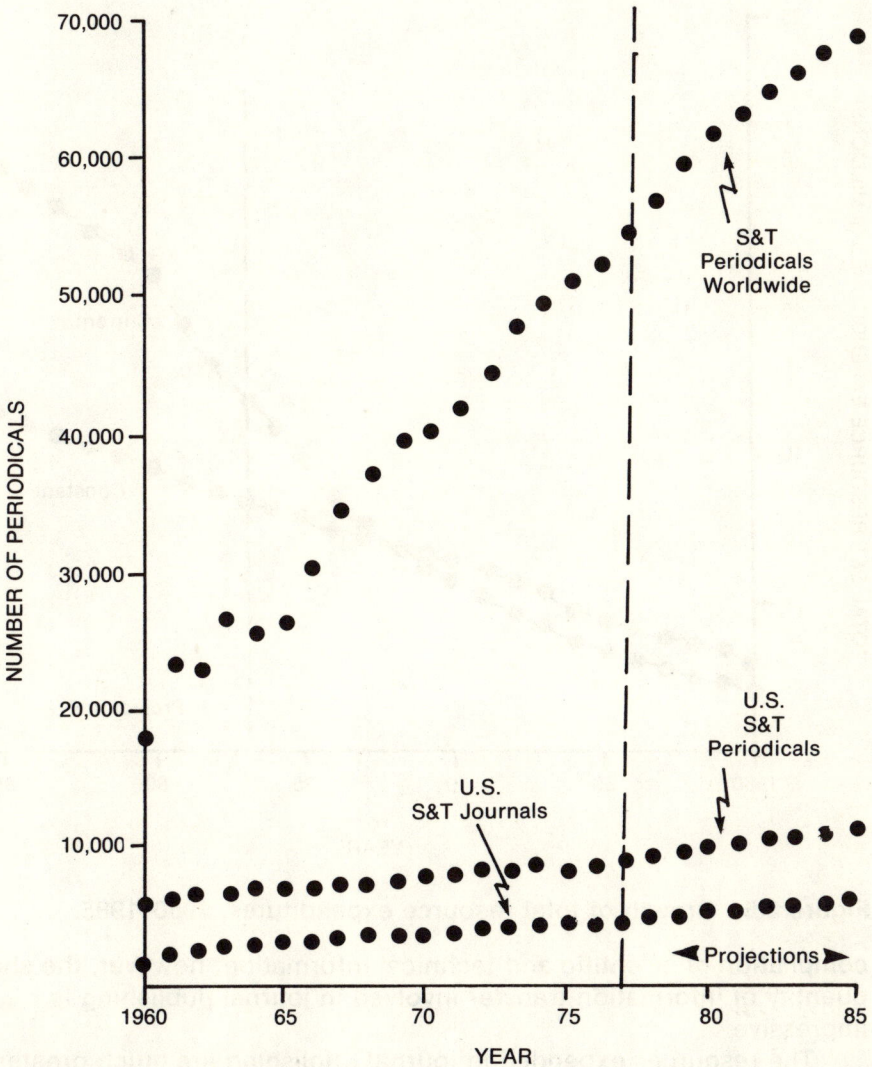

Figure 2.4 Growth of worldwide periodicals, U.S. periodicals, and U.S. journals, 1960–1985.

Much more activity is associated with journals than with the other forms of published literature. In the United States alone, over 2.4 billion copies of scientific and technical articles are distributed by publishers each year through subscriptions. This compares to approximately 15 million copies of books distributed. One can argue that books represent a more careful (and individually more comprehensive)

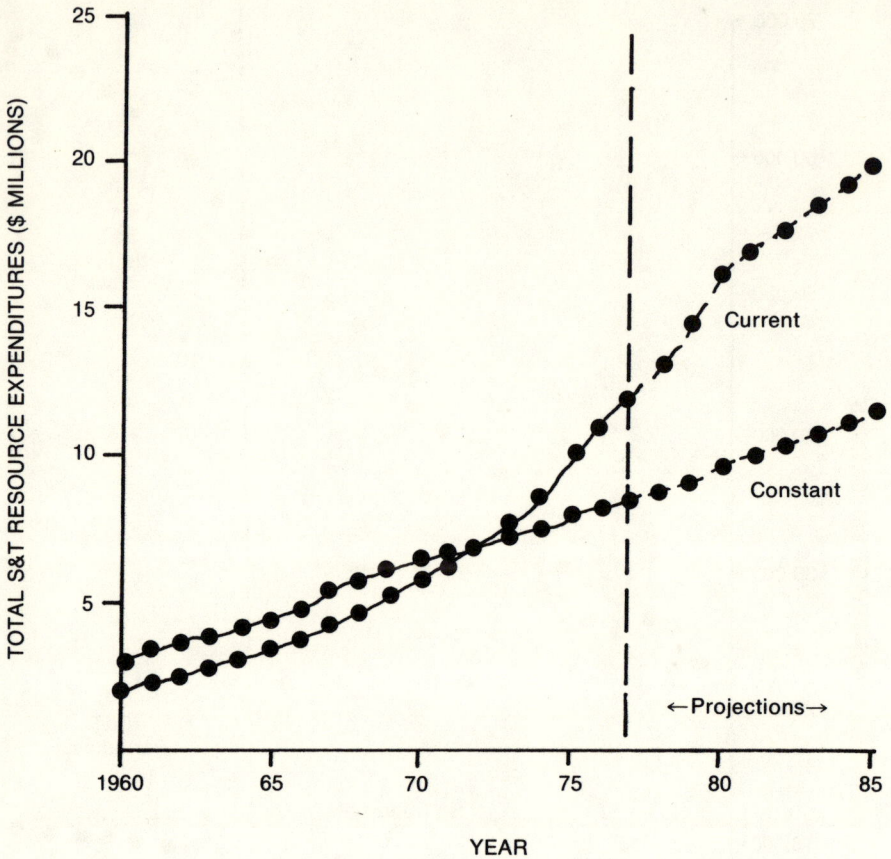

Figure 2.5   Growth of total resource expenditures, 1960–1985.

compilation of scientific and technical information; however, the sheer quantity of information transfer involved in journal publishing is impressive.

The resources expended in journal publishing are much greater than for other forms of the published literature. The total amount of resources expended on the scientific and technical literature was $12 billion in 1977. These expenditures include all activities associated with writing, publishing, distributing, library operations, secondary services, and reading the published literature. The growth of these expenditures, along with those attributed to journals, books, technical reports, and other forms, are shown in Figure 2.5. In 1977 an estimated $4.7 billion worth of resources were expended on scientific and technical journals, 61 percent of the total communication expenditures.

Resource expenditures for scientific and technical communication

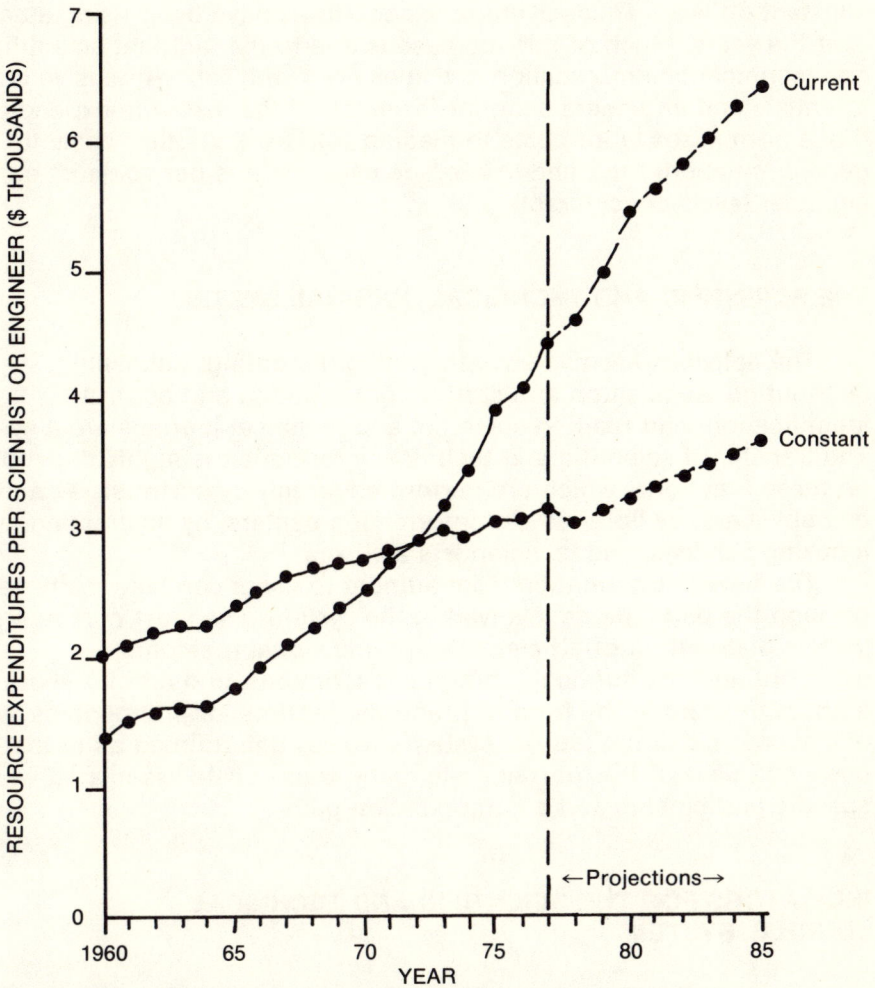

Figure 2.6 Resource expenditures per scientist or engineer, 1960–1985 (current and constant dollars).

and for journals and other periodicals represent a small but growing portion of the gross national product (GNP) and of research and development (R&D) expenditures in the United States. The ratios of scientific and technical communication expenditures were estimated to be 0.006 and 0.28 of 1977 GNP and R&D expenditures, respectively. These ratios, increasing in recent years, appear to have stabilized. Anticipated growth paralleling GNP more closely in the future suggests a relatively mature economic sector.

   Resource expenditures per scientist or engineer for scientific and technical communication are displayed in Figure 2.6 in current and constant dollars.* These average expenditures have been rising steadily over the years. Much of this increase is due to the fact that scientific and technical communication activities are highly labor, intensive and scientists and engineers account for much of the resource expenditures. If one normalizes labor costs to median salaries and other costs to general inflation, the trend in average expenditures per scientist or engineer levels considerably.

THE SCIENTIFIC AND TECHNICAL JOURNAL SYSTEM

The activities associated with writing, recording, publishing, distributing, acquisition and storing, organization and control, identification, and reading scientific and technical journals are a system. The transfer of scientific and technical information is highly dependent on these functions, which are performed largely by scientists as authors, by publishers, by libraries and information centers, by abstracting and indexing services, and by scientists as users.

The flow of information from authors to users can take many paths through the participants involved in the system. The most common ones involve distribution of articles through individual (personal) subscriptions, institutional (library) subscriptions, and articles sent from authors to users in the form of preprints, reprints, or photocopies. Much of our analysis of the journal system involves determining all of the costs and uses of the journal system and those costs associated with specific participants and communication paths.

INDICATORS FOR THE SCIENTIFIC AND TECHNICAL JOURNAL SYSTEM

There has been a great deal of speculation in recent years in the United States concerning the usefulness of scientific and technical journals. The emphasis on publishing in universities has certainly added some credence to the feeling that many useless articles are written simply to satisfy this need. In fact, only 6 percent of the population of scientists and engineers wrote the journal articles published in 1977 and 62 percent of these authors were employed by universities. The authors tend to be younger, with higher education than other scientists or engineers, indicating they may be trying to enhance their reputation through publishing. Although a high proportion of authors are from universities, most of them are primarily engaged in research (54

*Constant dollars are based on the 1972 GNP price deflator.

percent), much of which is funded by the government (57 percent of all articles). Furthermore, preparing articles, even though expensive ($2,300 per article), has the advantage of forcing the authors to clarify their research results since they will undergo peer review prior to and following publication. We also find that journal articles are frequently read (640 readings per article).

It is estimated that there were approximately 245 million individual readings of journal articles in 1977, most of them done with great care (39 percent) or at least with attention to the main points (57 percent). This amount of reading comes to about 640 readings per article, or 55,000 readings per journal, on the average. Scientists and engineers devote a substantial portion of their time to reading the journal literature, and this effort must be considered a rough assessment of their perception of the value of it. Based on our user survey, scientists spend an average of sixty-seven hours per year reading an average of eighty-eight journal articles. This ranges from forty hours per year for computer scientists to over two-hundred hours for mathematicians. In ten previous studies, it has been observed that scientists or engineers spent somewhere between thirty hours per year for engineers to over three-hundred hours per year for cancer researchers. The evidence of the large time devoted to reading journals is overwhelming.

An indication of value of journals is the purposes for which articles were read in 1977. The scientists and engineers who responded to our user survey indicated that nearly all of them applied the article findings to a current project (44 percent for research findings and 45 percent for research methods) or a research proposal (17 percent); 16 percent used it in preparation of an article, book, review, or report; 18 percent used it in preparing a lecture or presentation; and 6 percent used it for planning, budgeting, and management of research.

The total cost of reading articles is estimated to be $2.3 billion, so the average cost per reading is $9.60. Scientists and engineers also devote a great deal of time to preparing journal articles. Of the 382,000 articles published in the United States, about 294,000 are estimated to be written by U.S. authors. The U.S. authors also wrote an estimated 70,300 articles that were published in non-U.S. journals. Since the authors and coauthors spend an average of eighty-two hours preparing articles, they expended a total of 30 million hours over all the journal articles written in 1977. Including other support costs, the total costs associated with authorship come to $0.9 billion, or $3.60 per reading. The incentives for writing are certainly different from those for reading; however, the volume of writing is one more indicator of the importance of the journal literature to the scientific and technical community.

Publishers play a crucial role in scientific and technical communication. They serve as a focal point for gathering research

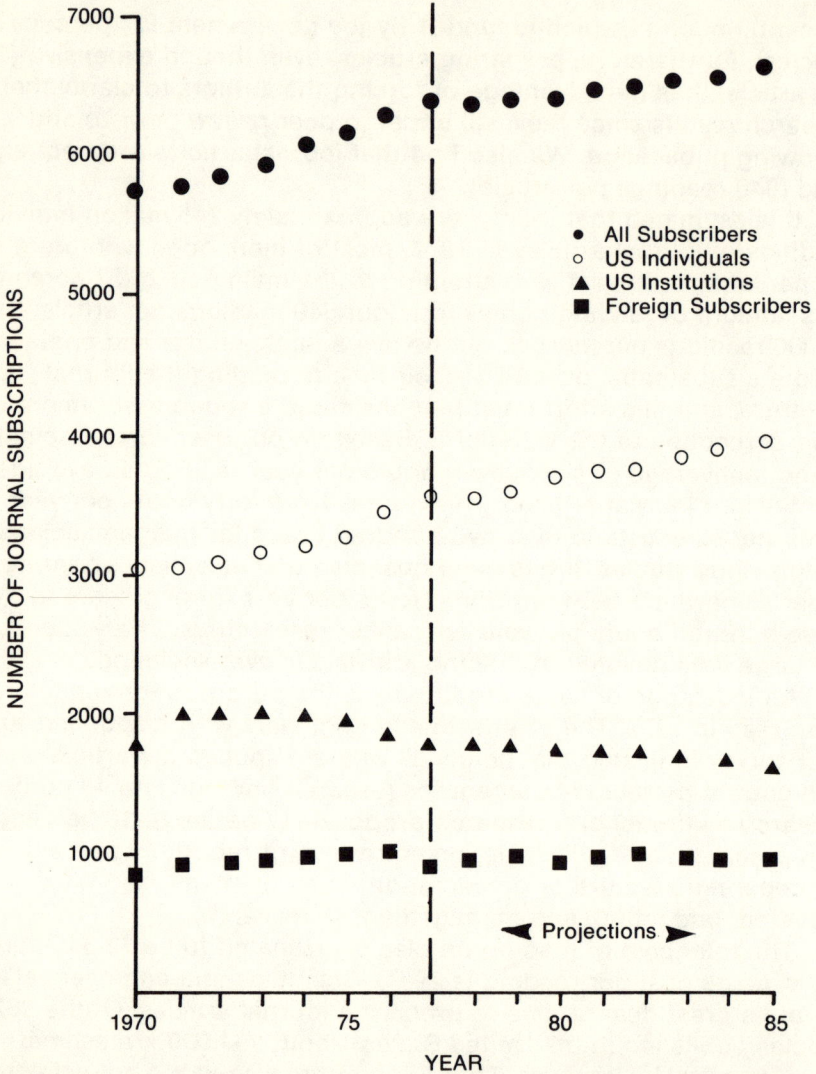

Figure 2.7 Average number of journal subscriptions by type of subscriber, 1970–1985.

manuscripts, reviewing and editing them, preparing them for distribution, establishing an audience for the information, and reproducing and distributing the information to this audience. This is no small role; nearly 382,000 articles are involved and over 2 billion copies of these articles were distributed by publishers in 1977 through an estimated 4,447 journals published in the United States. The average

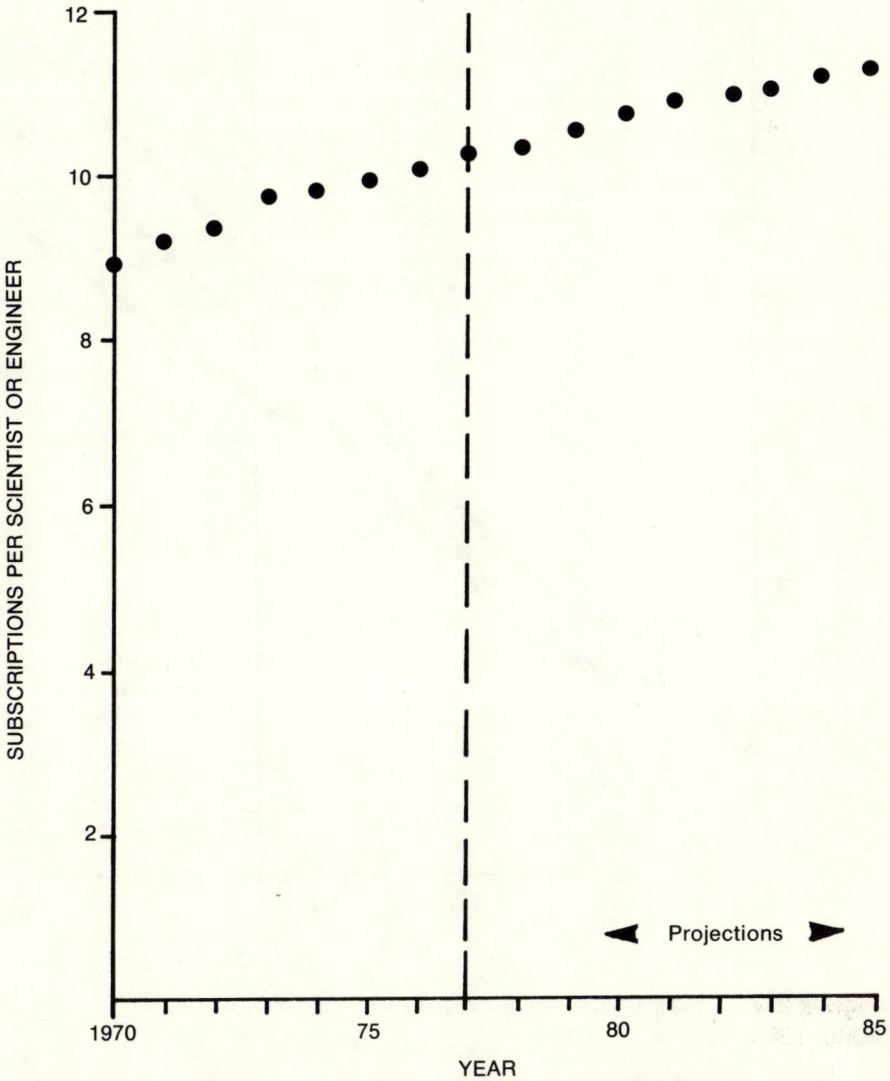

Figure 2.8   Number of subscriptions per scientist or engineer, 1970–1985.

number of subscriptions for these journals is estimated to be 6,327. The average number of subscriptions has been increasing steadily since 1970, the first year for which reasonably good data are available (Figure 2.7). Much of this growth is due to the increase in number of scientists and engineers. However, even the number of subscriptions per scientist or engineer is growing slightly (Figure 2.8). In 1970 the

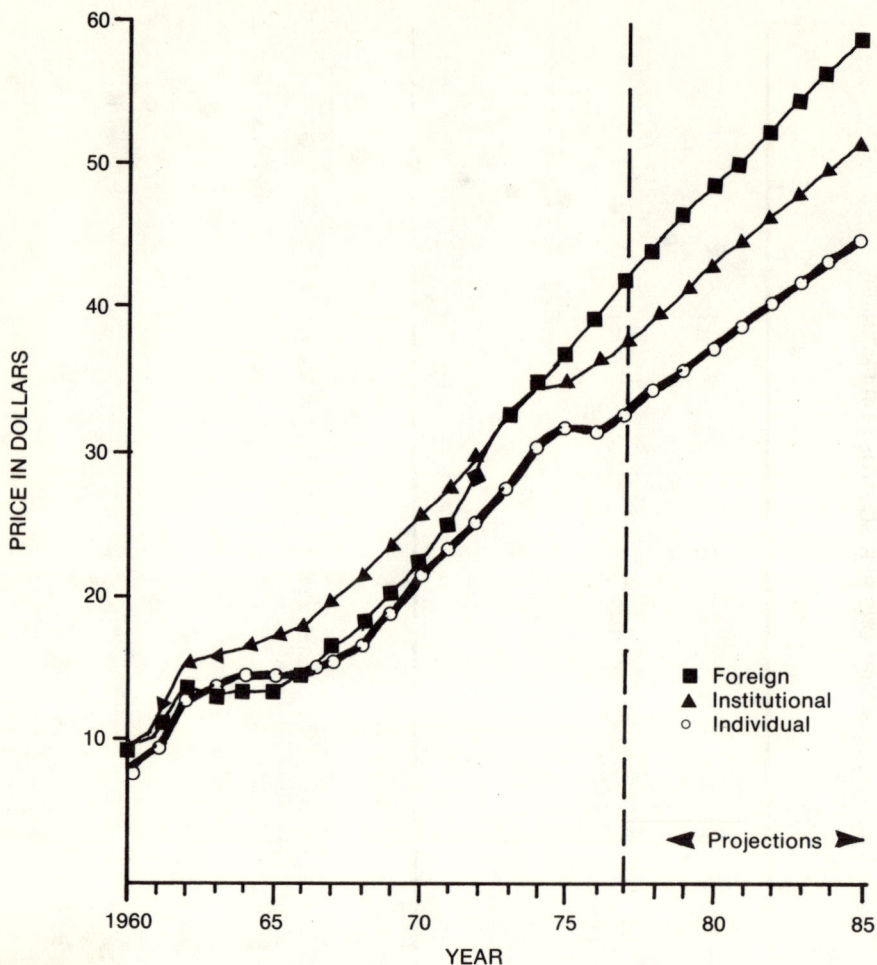

Figure 2.9   Journal subscription price in current dollars, 1960–1985.

number was 8.8 subscriptions per scientist or engineer; this number grew to 10.1 in 1977. Most of the growth of subscriptions comes from U.S. individual subscribers; the number of U.S. institutional subscribers (thought to be mostly libraries) is decreasing.

The overall increase in subscriptions has taken place even though journal prices have increased dramatically over the years. The average journal prices are given in current and constant dollars in Figures 2.9 and 2.10. The price per journal varies among the types of subscribers. In

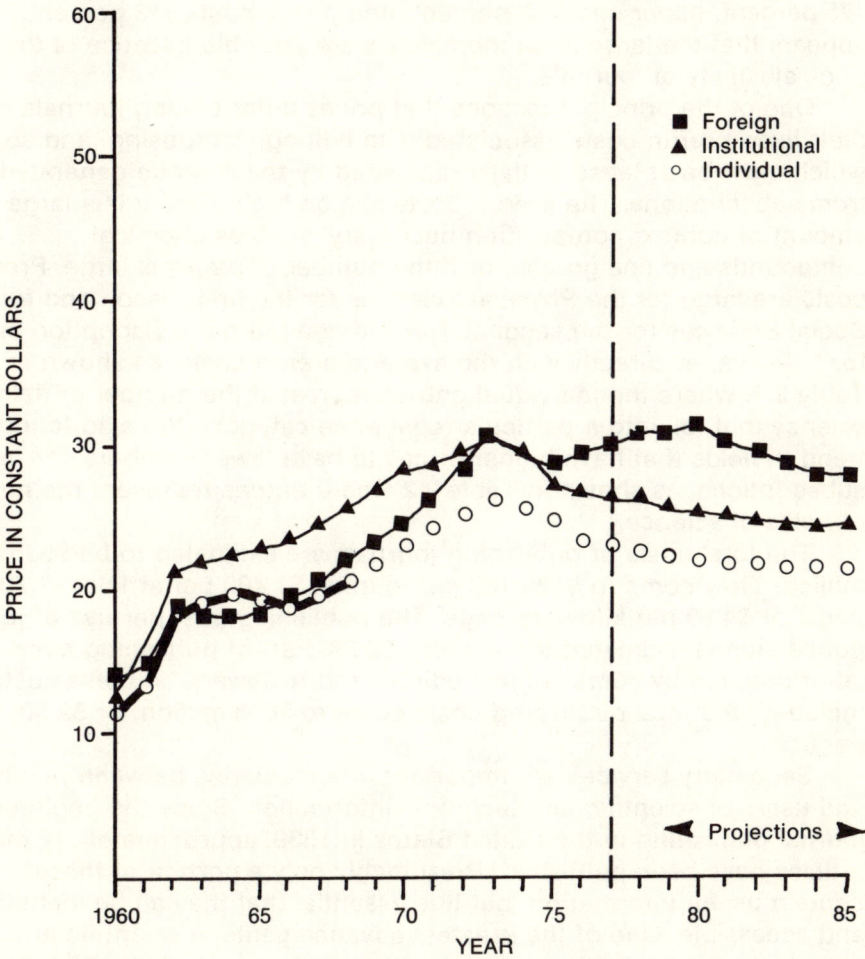

Figure 2.10 Journal subscription price in constant dollars, 1960–1985.

1977 the price to U.S. individuals was $32.70, to U.S. institutions $37.70, and to foreign subscribers $42.05. The rate of growth of prices to the three types of subscribers has not differed very much (317 percent individual, 306 percent institutional, and 348 percent foreign from 1960 to 1977).

The price increases partially reflect the rapid rise in publishing costs. Between 1960 and 1977, for example, editors' salaries rose 142 percent, typesetting costs rose 179 percent, printing costs

175 percent, paper costs 52 percent, and postal costs 113 percent. It appears that the large price increases were possible because of the price elasticity of journals.

One of the principal reasons that prices differ among journals is their fixed prerun costs associated with editing, composing, and so on, which must be at least partially recovered by the revenue generated from subscriptions. The prerun costs can be high if there is a large amount of nontext composition necessary, such as chemical compounds and line graphs, or if the number of pages is large. Prerun costs are large for the Physical Sciences for the first reason and for the Social Sciences for the second. The average journal subscription price for fields varies directly with the average prerun costs, as shown in Table 2.1, where the individual entries represent the number of fields of science that fall into a particular cost/price category. We also found a trend in fields that have higher prices to have lower numbers of subscriptions, as shown in Table 2.2 where entries represent the number of fields of science.

The total costs of publishing journals are estimated to be $535 million. They come to $120,400 per journal, $1,400 per article, $1.70 per page, or $4.90 per kiloword page. The publishing cost per use of journal publication is estimated to be about $2.20. Part of publishing costs are labor donated by some subject editors and reviewers.* If these costs are included, the total publishing costs come to $618 million, or $2.50 per use.

Secondary services are important intermediaries between publishers and users of scientific and technical information. Since the beginning of journal publishing in the United States in 1839, approximately 12 million articles have been published. Presumably only a portion of these contain useful information, but it is essential that they all be identifiable and accessible. One of the greatest advancements in scientific and technical communication has been the development of identification and location products and services. The number of items processed annually by abstracting and indexing services is estimated to be 3.2 million, up from 1.1 million in 1960. In 1975 there were about 179 data bases in the United States and another 122 in other countries. The number of computer data-base searches has increased dramatically since 1968, from about 10,000 to an estimated 1.8 million in 1977.

Our user survey revealed that articles are frequently identified from citations in printed indexes. For example, readers in the Physical

*Throughout the book we refer to peer reviewers or referees of article manuscripts as reviewers (not to be confused with book reviewers).

Table 2.1 Number of fields of science by average journal prerun costs and average journal subscription price, 1977.

Average Prerun Costs	Average Subscription Price			
	$10–30	$31–50	$51–70	Over $70
Over $100,000	0	0	0	1
$81,000–100,000	0	1	1	0
$61,000–80,000	0	2	0	0
$40,000–60,000	2	0	0	0

Table 2.2 Number of fields of science by average number of journal subscriptions and average journal subscription price, 1977.

Average Number of Subscriptions	Average Subscription Price			
	$10–30	$31–50	$51–70	Over $70
Over 10,000	1	2	0	0
7,000–10,000	0	0	0	0
4,000–10,000	1	0	1	1
Under 4,000	0	0	1	1

Sciences, Computer Sciences, and Engineering all had about one-third of their read articles identified in this manner. The Environmental Sciences and Other Sciences had very few read articles identified in this way. The other fields had between 10 and 15 percent of their read articles identified from citations in printed indexes. An estimated 2 percent of the articles read in the Life Sciences were identified using computer searches.

Libraries acquire, store, and maintain scientific and technical journals and also utilize the data-base products and services to assist users to find needed information. The number of libraries has grown somewhat since 1960 (Figure 2.11). The acquisition of journals has also grown, from 5.1 million subscriptions in 1965 to 8.3 million in 1977. However, the prices of journals have increased substantially faster than library budgets, which has probably had some dampening effect on library acquisitions. In fact, the average number of institutional (library) subscriptions per journal and number of library subscriptions per scientist or engineer has actually decreased.

The total cost of library activities associated with scientific and technical journals was estimated to be $433 million in 1977, about $1.80

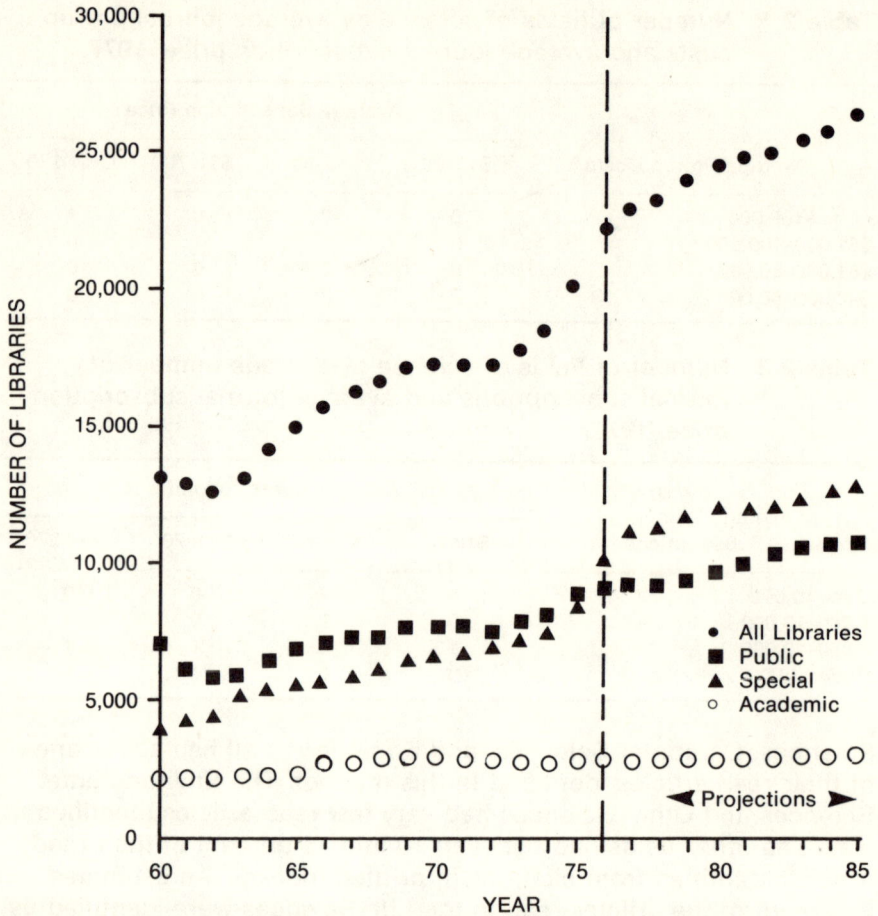

Figure 2.11 Growth of U.S. libraries by type, 1960–1985.

per use. The estimated cost of secondary services, including abstracting and indexing and data-base searching, comes to $152 million, about $0.60 per use.

Most of the readings occur from journal subscriptions that are sent to individuals (179 million); library subscriptions are read second most frequently (36 million) and author preprints, reprints, and photocopies next (30 million). Since the total number of individual subscriptions is 15.8 million, the average number of articles distributed by this means is 7.6 article copies per reading. With the 8.3 million library subscriptions, this number comes to 19.7 articles distributed to libraries per reading. However, the cost per use of the three principal paths varies substantially: $16.20 for individual subscriptions, $33.70 for library subscriptions, and $20.00 for author preprints, reprints, or photocopies. The average cost per use over all paths is $18.60.

Based on these limited data, it appears that the least cost-effective path is the one involving libraries. However, such an assessment is not valid because the articles obtained in libraries differ somewhat from those read in individual subscriptions. For example, they are much older than the ones read from individual subscriptions. The average half-life of articles read from library subscriptions is about three times that observed in individual subscriptions.*

SOME OBSERVED WEAKNESSES IN THE JOURNAL SYSTEM

Despite mostly favorable indicators concerning scientific and technical journal publishing, there are some ominous signs. Journal resource expenditures are rising faster than R&D expenditures (584 and 153 percent, respectively, from 1960 to 1975). Since there is a substantial overlap between the two, there may be a tightening of journal expenditures as they come into stronger competition for R&D funds. This conflict is eased somewhat because much of the time spent by scientists in writing, editing, reviewing, and reading journal articles is performed during nonworking (unpaid) hours.† Certainly there has been a correlation in the past between the fact that most authors come from academic institutions and the fact that they had more time available to write for publications.

Three factors appear to make continued coverage of the journal literature very difficult for scientists and engineers. First, the number of articles appearing in U.S. publications is increasing slightly faster than the number of scientists and engineers. If the readings per article remain constant, then scientists and engineers must read more current articles per year on the average. In fact, if these trends continue, the number would increase from seventy-four articles per year in 1965 to ninety-one articles per year in 1985. Second, the accumulation of journal literature over the years makes continued coverage even harder. Roughly 200,000 journal volumes (over 12 million articles) have been published in the United States since 1839. In the next ten years this

---

*Half-life of articles refers to the age at which one-half of the readings occur prior to and after that time.

†An economic value is assigned to this labor in order to evaluate the total resources engaged. The opportunity cost to the R&D system, however, is low because very little additional research would be done in the time not spent producing journal articles.

number could increase by another 4 or 5 million articles. Third, the even faster growth and accumulation of the non-U.S. journal literature exacerbates these problems.

Since most activities involved in the journal system are labor intensive, journal costs are rising steadily, even in constant dollars. Some signs of this economic pressure on the system are beginning to surface in attempts of one participant to shift cost (or cost recovery) to another participant. One example of this is in page charges and related fees. Publishers attempt to shift some of their publishing cost burden to authors by asking them to pay a fee according to the length of the article, submission fees, reprint fees, or some other form of fee for placing their articles into the journal system. Currently, approximately one-half of the articles published are covered by such fees. These costs are usually passed on to an employer (32 percent) or the federal government (68 percent),* but this does not alter the fact that an attempt is made by publishers to shift these costs.

Another indication of economic pressures is the bitter controversy displayed during the debate over the new copyright law. Publishers incur a large cost before any copies are reproduced. Evidence suggests that institutional (library) subscriptions are decreasing slightly. In order to recover some of the lost income, publishers are requesting royalty payments for a portion of the photocopies made of their articles by libraries and others. Thus they are attempting to shift more of the cost burden to the libraries. Another method to achieve this aim is through increased library subscription prices. Prices to libraries have gone up 306 percent from 1960 to 1977, an increase of 197 percent in constant dollars. A question is whether subscriptions began to drop, thereby requiring higher prices to recover large fixed costs, or vice-versa.

Libraries also are at fault since they subscribe to more journals than economically necessary.[5] They could more often obtain copies of articles from infrequently used journals through interlibrary loans. Thus some libraries are substituting interlibrary loan costs, which are shared by the lending and borrowing libraries, for journal acquisition and maintenance costs. This procedure creates an economic burden on the lending library. Accordingly some libraries have instituted charges for interlibrary loans, and there is continued discussion of whether libraries should charge users for individual information products and services.† This appears unlikely to happen in the near future, but its serious contemplation is a further indication of the economic squeeze on

*Sometimes the employer is the federal government.
†In contrast to the well-established practice of student library fees, industrial departments' pro-rata cost allocations and other forms of undifferentiated subscriptions to library services.

libraries. For academic libraries part of this economic problem is due to the fact that recent enrollments not having kept pace with other growth factors because of decreasing birthrates in the 1960s.

There is some evidence[6,7] that libraries acquire many journals that are little used by patrons. This places librarians in a dilemma: they feel that libraries must continue the archival role in order to serve their patrons' literature needs; yet maintaining the archive competes directly for funds that could enhance other services, such as reference capabilities. It has been estimated[8] that it is less expensive to borrow copies of journal articles through interlibrary loans than to purchase the pertinent journals if there are fewer than five to ten uses for one year's publication of that journal. Several library studies have shown that a high proportion of journals fall into the category of having fewer than five uses. One[9] estimated that 25 percent of the journals in a physics library have fewer than five uses and 43 percent fewer than ten uses per annual volume.

Clearly some libraries have recognized this economic trade-off.[10] The number of interlibrary loans has increased to 5.6 million copies of scientific and technical articles published in the United States. Interlibrary loans are expensive compared to average cost to use a library subscription copy (albeit less expensive compared to purchased journals under low frequency of use conditions). It is estimated to cost $20 per loan on the average with $11.60 cost to the borrowing library plus $8.40 cost to the lending library.

The new copyright law is partially directed to this practice of interlibrary lending.[11] The guidelines[12] concerning interlibrary loans promulgated by the Commission on New Technological Uses of Copyrighted Works (CONTU) suggest that the burden of royalty payment is on the borrowing library. However, the borrowing library is not required under these guidelines to pay royalty fees for articles that are photocopied under fair use (for example, to replace mutilated copies or for classroom use), or for articles published more than five years prior to when copies are made. Also they are not required to pay royalty fees for fewer than six copies made for the borrowing library of articles in one annual volume of a journal.

One of the purposes of the copyright law is to protect the property rights of the authors and the journal publishers, if the latter acquire these rights. Obviously publishers are concerned that they not lose a large part of their circulation. It is estimated that 34 percent of the domestic subscriptions in 1977 (and 16 percent of foreign subscriptions) were to institutions or libraries. Thus a large reduction in this segment of their market could hurt them badly. The fundamental problem is that journal publishing is characterized by large, fixed costs associated with editing, composing, and so on. Thus average costs do not vary much by number of subscriptions. The average fixed costs are $72,800 per journal per year. If the number of subscriptions goes down, less of the $72,800 fixed costs will be recovered from the subscription revenue or the subscription price must go up; this in turn would mean that some more subscriptions will be lost. For this reason, publishers would like to have another means of recovery for these prerun costs.

Journals with a low number of subscriptions are particularly vulnerable to a small change in the number of subscriptions. A small subscription journal (fewer than 3,000 circulation) must on the average recover $52 per subscription for the fixed costs compared to $13 for medium subscription journals (3,000–10,000) and $3 for large subscription journals (over 10,000 circulation). This is a dramatic difference that partially accounts for the fact that the average prices of small subscription journals are double those of large subscription journals.

Since the price to libraries of small subscription journals is greater than the large subscription journals, the breakdown point in number of uses for library purchasing versus borrowing is higher for the small subscription journals. Thus they are more subject to cancellation than the large subscription journals. There is some evidence[13] that articles of small subscription journals are borrowed slightly more often than those of large subscription journals. This can be more damaging to the former because a higher proportion of their subscriptions are from libraries (45 percent compared to 26 percent for large journals). If libraries began to operate in a strictly economically rational manner, evidence shows that nearly 30 percent of their library subscriptions would be canceled. Either the publishers of the small journals would have that much less revenue or they would have to increase their prices even further, in which case there would be a further trend toward less circulation and higher prices.

Additional revenue from royalty payments may not be sufficient in comparison to lost subscription income because the CONTU guidelines limit the royalty payments to journals having more than five loans by the borrowing library. But that is about the breakeven point at which libraries should borrow rather than purchase journals. We believe that the new copyright law will have little effect overall on the journal system because of the relatively small number of photocopies involved, but it could significantly affect small journals. Without the ability to collect funds, they could be left virtually unprotected.

Small subscription journals have another economic disadvantage. Since they have low circulation, they have little or no revenue from advertisements. Average advertising income for small subscription journals is about $500 compared to $20,000 for large subscription journals. One might argue that small subscription journals are not worthwhile and that the marketplace should determine whether they survive. Certainly some information published in small journals is not very important to our fund of knowledge. Some information has a small audience, with perhaps fewer than one hundred scientists having a possible interest in it. Yet we should have some way of communicating the information and preserving it for later use. Gregor Mendel published the basic law of genetics in what was even in his day a small journal.

NEW TECHNOLOGY FOR THE JOURNAL SYSTEM

One of the indicators of inefficiency in the current journal system is the fact that about ten copies of articles are distributed for each use. Even though the overall cost is a reasonable $19 per use, there are still some wasted funds in the sense that many articles are distributed but not used. The future seems to hold some promise for improving the system through some developments that are occurring in the scientific and technical communication community and some outside that could have profound impact on the journal system. A recent study performed for NSF[14] found that electronic processes could make a great difference in future communications.

Electronic word-processing and text-editing systems have already achieved widespread use in authorship. Word-processing systems use editing typewriters with magnetic digital storage. Text-editing systems range from terminals that tie into a local computer to intelligent terminals with self-contained memory and microprocessor-based computing functions. These systems yield improvements in editorial quality of manuscripts and provide faster and much more reliable operations. Their most direct advantage comes from the substantial reduction of secretarial labor costs, with the cost trade-off of increased investment in equipment. Heavy competition in this area is resulting in increased sophistication and flexibility as well as reduced costs. Options such as editing and spelling routines, special print stations, message service, and, perhaps most importantly, photocompositor interfaces are available. In the long run, manuscripts will be transmitted to publishers in digital form, and articles will be edited and composed by electronic processes, substantially reducing prerun publishing costs.

Electronic processing is being used extensively by publishers for editing, redaction, and composition. The possibility of direct input from authors to publishers in digital form would increase the advantage of such processing since the publisher's cost of keyboarding can be eliminated. Optical character recognition (OCR) might also be used increasingly for direct input. However, its technology and costs have improved much more slowly than many anticipated. Composition is being increasingly integrated into general text processing in computer-based systems, which were originally designed primarily for editing text. The magnetic digital record can also be stripped of operating codes, provided with typesetting control characters, and input into an electronic composer. This type of integrated system is often used by newspaper publishers.

In the future, publishers will have enormous flexibility in the form and mode of distribution of articles made from master images (formatted article text). Electronic technology has progressed to the point where articles can be individually printed directly from computer output. Nonimpact printing has resolved the problem of computer output speed. Ink jet, electrophotographic, and electrostatic systems all hold substantial promise for scientific and technical publishing.

Computer master images can also be transformed into microform or video discs. However, microform still has many disadvantages: too many formats, reduction ratios, and retrieval coding schemes; poor quality film images at the reader and paper blow-back prints; readers that are expensive, hard to use, and bulky; inadequate user environments in libraries; and difficulty in obtaining paper prints. The greatest promise for microform seems to be with new libraries that wish to build up an archival file in certain areas inexpensively. To date, however, the savings in reproduction (runoff) are not great enough to overcome the large prerun costs. Prices cannot be sufficiently reduced to induce libraries or scientists to purchase these forms of distribution. The new video disc technology also holds interesting promise for distribution and storage. We feel that there are potential weaknesses here also, since their images currently are not adequate for displaying a full normal-sized printed page on most inexpensive viewing devices. If these weaknesses can be overcome, there is a huge potential in the use of video discs in libraries, where an entire library collection can be stored in a few cubic feet of space, revolutionizing library storage and operations.

One of the areas in which new technology has been directly applicable to scientific and technical communication is in library operations and services. The most prominent of these are on-line (and off-line) bibliographic searches, automated circulation, cataloging of books, and interlibrary loan processes. Technology outside scientific and technical communication that might make the greatest impact on the storage function is in mass storage memories. If a National Periodicals Center comes into being, one component might be digital storage of articles input directly from publishers. Thus far, the most likely such system is magnetic tape media segmented for mechanical handling. However, there are some output problems associated with queuing delays, and the output costs are prohibitively high. Other barriers to mass electronic storage are the needs for sophisticated indexing and control and for access software.

Most telecommunication currently is by voice grade telephone lines. Value-added networks have provided a potentially low-cost telecommunication capability for all participants in scientific and technical communication. This potential is enhanced by the availability of minicomputers and intelligent terminals, which permit rapid transmittal of messages that can then be buffered. Furthermore new technology in long-line communication, such as fiber optics, digital transmission and switching equipment, and communication satellites, could yield substantial decreases in cost if the capital cost of this technology can be shared by enough users.

These technological advances, which were developed largely independently of scientific and technical communication, provide all of the parts for a comprehensive electronic alternative to paper-based journal publishing. Such a system would provide enormous flexibility to the journal system. Individual articles could be distributed in the manner most economically advantageous. Highly read articles could still be distributed in paper form, while infrequently read articles could be requested and quickly received by telecommunication when needed. Here the trade-off is that resources formerly wasted in printing, mailing, and storage are applied to better identification and retrieval of information, thus reducing cost, improving quality, and increasing efficiency. We believe that better systems integration will yield more emphasis on quality of article content, that fewer articles will be repeated over and over again for updates or for different journals, and that the new systems will provide better access to and retrieval of information needed in multidisciplinary research.

In a comprehensive electronic alternative, articles will be prepared by authors using sophisticated text-editing systems. Article preparation may include joint text writing through teleconferencing systems in which immediate peer review is possible, comments are made, and specific research questions can be answered. Many of the citations used in the article will come from on-line bibliographic searches. When citations are identified they can be immediately retrieved by telecommunication in full text on cathode ray tube (CRT), or in paper form. The digital form of the unreviewed manuscript will be directly transmitted electronically to a publisher. The publisher will electronically transmit the manuscript to a subject editor, who will read the text by CRT or printout and make electronic notes concerning editorial and content quality. The subject editor may choose an appropriate reviewer by using a computer program that matches the profile of potential reviewers with the topics covered in the article. Other computer-stored information will also be used to help screen reviewers such as by affiliation and relationship to the authors, status of the most recent review, frequency of reviews, timeliness of response of previous reviews, and quality of reviews. The reviewers will respond to editors and editors in turn to authors by telecommunication, similar to current teleconferencing processes. Publishers and editors can also use electronic processes for business purposes, address listings, and so on.

An accepted article will be subject to redaction on a text-editing terminal, and the computer-based text will be output in several forms, including full text and bibliographic citation. The bibliographic form will be transmitted electronically and used directly by search services, or input to abstracting and indexing services to be analyzed and processed further. The full text will be sent electronically to some individual scientists designated by the author or by request to scientists based on its topic, author, or some other bibliographic identifier. In some instances articles will be telecommunicated through a selective dissemination of information-like system. The scientists may receive the text on their own personal terminals or on their libraries' terminals. Articles will be sent directly by telecommunication to a National Periodicals System based on a central archive. The system will await telecommunicated requests for copies of articles and will respond with full text telecommunicated to the requestors.

The electronic processes provide a great deal of flexibility of output, which can enhance reading and assimilation of the information. For example, end users could request alternative formats of the text that would suit their particular needs—for example, for rapid scanning or in-depth reading. Rapid scanning can be facilitated by highlighting certain elements of text, narrowing column widths, or widening space between lines. Human factors considerations can also help in-depth reading through other alternative format structures. Electronic processes can aid in combining mathematical formulas, data presentations, and text in ways that meet alternative needs.

This comprehensive electronic alternative system is highly desirable and currently achievable. We believe that a majority of articles will be handled by at least some electronic processes throughout all the system but that not all articles will be incorporated into a comprehensive electronic alternative system like that described. Some articles will be processed electronically in different ways depending on the electronic capabilities of the senders and receivers.

Technological constraints do exist. Standards must be set for word-processing and text-editing output so that publishers can receive it and easily convert it to the appropriate format. One major problem is treating nontextual material—tables, line graphs, photographs, mathematical formulas, and chemical compounds. Graphics can be electronically handled now, but the economics are not practical for the substantial amount of graphics found in articles published in the fields of Physical Sciences, Engineering, and Life Sciences. Another requirement is lower cost of telecommunication. Unfortunately there is some indication that future Federal Communications Commission rulings could damage this prospect.[15] Sending and receiving equipment must be sufficiently sophisticated to permit rapid communication in order to reduce costs. Mass storage devices now available could economically store nearly all current literature, but the costs of input and output may be unacceptable unless some breakthroughs in this area are made.

There are also some major participant constraints in adoption and use of an electronic journal system. One of the principal ones to any alternative communication system is the lack of incentive for the communication participants to change. For example, authors are said to publish partially for prestige and recognition, which results in professional advancement; therefore any alternative communication system must meet this perceived need. Many publishers lack a financial incentive for drastically deviating from the current journal publishing practices. While many book and small journal publishers appear to have financial problems, most large publishers are doing very well financially. They earn a comfortable margin on income, and they require much less capital to publish journals than books since the income from subscriptions is received before most costs are incurred. The return on investment for journal publishing is therefore favorable. Substitution of royalty payments for subscription income will lessen this advantage; since photocopying takes place over a long time, the current value of royalty income is less than the value of an equal amount of subscription income. Thus any new publishing systems must incorporate some financial incentives or publishers will be unlikely to want to change.

Scientists as users also present some barriers to new systems that directly affect their behavior. If an alternative journal publishing and distribution system involves direct on-line communication, some incentives to use it must be provided to scientists. We believe that in the future, new scientists who have been trained on terminals in high schools and universities will find it unacceptable not to have these facilities available for analysis, text processing, search, retrieval, and other forms of communication.

Libraries have little incentive to modify their mode of operating unless their patrons and funders desire such change. While many libraries currently are automating for cataloging and internal record keeping, they still require motivation to change their procedures in dealing with scientists. Here again, some outside incentive will probably be necessary.

These problems to be solved are technological, economical, and institutional. We believe that the analyses in this book illustrate many of the important problems so that policy decisions can be based on sound information.

REFERENCES

1. American Psychological Association, *Reports of the American Psychological Association's Project on Scientific Information Exchange in Psychology,* Volume 1 (Washington, D.C.: American Psychological Association, 1963), vol. 1.
2. W. D. Garvey and S. D. Gottfredson, "Scientific Communications as an Interactive Social Process," *International Forum on Information Documentation* (2(1):9–16 (1977).
3. W. D. Garvey, N. Lin, and K. Tomita, "Research Studies in Patterns of Scientific Communication. III. Information Exchange Processes Associated with the Production of Journal Articles," *Information Storage and Retrieval* 8(5):207–221 (1972).
4. D. S. Price, *Little Science Big Science* (New York: Columbia University Press, 1963).
5. *A. Kent, K. L. Montgomery, J. Cohen, J. G. Williams, S. Bulick, R. Flynn, W. N. Sabor, and J. R. Kern, A Cost-Benefit Model of Some Critical Library Operations in Terms of Use of Materials* (Pittsburgh, Pennsylvania: University of Pittsburgh, 1978).
6. Ibid.
7. C.-C. Chen, "The Use Pattern of Physics Journals in a Large Academic Research Library," *Journal of the American Society for Information Sciences* 23(4):254–265 (1972).
8. V. E. Palmour, M. C. Bellassai, and R. R. V. Wiederkehr, *Costs of Owning, Borrowing, and Disposing of Periodical Publications* (Arlington, Virginia: Public Research Institute, 1977).
9. Chen, "Use Pattern of Physics Journals."
10. King Research, *Library Photocopying in the United States: With Implications for the Development of a Copright Royalty Payment Mechanism* (Washington, D.C.: Government Printing Office, 1977).

11. Public Law 94-553, *A Copyright Act* (Washington, D.C.: 94th Congress, October 19, 1976).
12. U.S. Congress, "Joint Explanatory Statement of the Committee of Conference," *Congressional Record House,* pp. H11 728-H11 729 (September 19, 1976).
13. King Research, *Library Photocopying.*
14. D. W. King and N. K. Roderer, *Systems Analysis of Scientific and Technical Communication in the United States: The Electronic Alternative to Communication Through Paper-Based Journals* (Rockville, Maryland: King Research, 1978).
15. M. Gerla, "New Line Tariffs and Their Impact on Network Design," *AFIPS Conference Proceedings* 43:577–582 (1974).

Chapter 3

Authorship of Scientific and Technical Journals

DESCRIPTION OF THE SCIENTIFIC AND TECHNICAL COMMUNITY

The community of U.S. scientists and engineers is the source of both producers and users of journal articles. There were estimated to be approximately 2.8 million scientists and engineers in the United States in 1977.[1] This number has increased steadily since 1960 and is expected to continue to grow through 1985 (see Figure 3.1). The average annual rate of growth from 1965 to 1977 was 3.3 percent and the average rate to 1985 is estimated to be 3.0 percent. Growth rates among the individual fields of science vary over the years, as do the proportions of the whole accounted for by the individual fields (see Table 3.1). About half of the entire scientific community was represented by engineers in 1977 (1,383,000). The smallest group is found in Environmental Sciences (89,000). The fastest-growing fields by far are Computer Sciences, which increased 519 percent between 1965 and 1977, the Social Sciences (388 percent), and Psychology (268 percent). The Physical Sciences represent the slowest growth during this period of time (11 percent). In fact, from 1970 to 1975 there was actually a slight decrease in number of scientists in this group. There was a similar lack of growth in Mathematics during the same time period.

The proportion of scientists and engineers actually employed in 1977 is not known. However, data are available for the proportion employed in 1974 by field of science. According to the National Science Foundation's Manpower Characteristics System,[2] about 84 percent were employed, and the labor participation rate was 85 percent.* These data are given by field of science in Table 3.2. The greatest participation rate is found in Engineering, the lowest rate in Social Sciences.

*Labor participation is defined as being in the labor force. Included are those employed as well as those unemployed and seeking employment.

49

Table 3.1 Number of scientists and engineers, by field of science, 1965–1985 (in thousands).

Year	Physical Sciences (No.)	(%)	Mathematics (No.)	(%)	Computer Sciences (No.)	(%)	Environmental Sciences (No.)	(%)	Engineering (No.)	(%)	Life Sciences (No.)	(%)	Psychology (No.)	(%)	Social Sciences (No.)	(%)	All Fields (No.)	(%)
1965	253		52		31		50		1,182		212		37		50		1,867	
1970	299	18	115	121	134	332	63	26	1,338	13	268	26	71	92	97	94	2,385	28
1975	275	8	108	6	171	28	83	32	1,340	<1	308	15	119	68	231	138	2,635	10
1977	287		112		192		90		1,378		323		136		244		2,762	
1980	305	11	121	13	228	33	101	22	1,451	8	347	13	165	39	263	14	2,981	13
1985	335	10	136	15	284	25	120	19	1,578	9	388	12	212	28	296	13	3,349	12

Note: The percentage changes are for the previous five-year period.

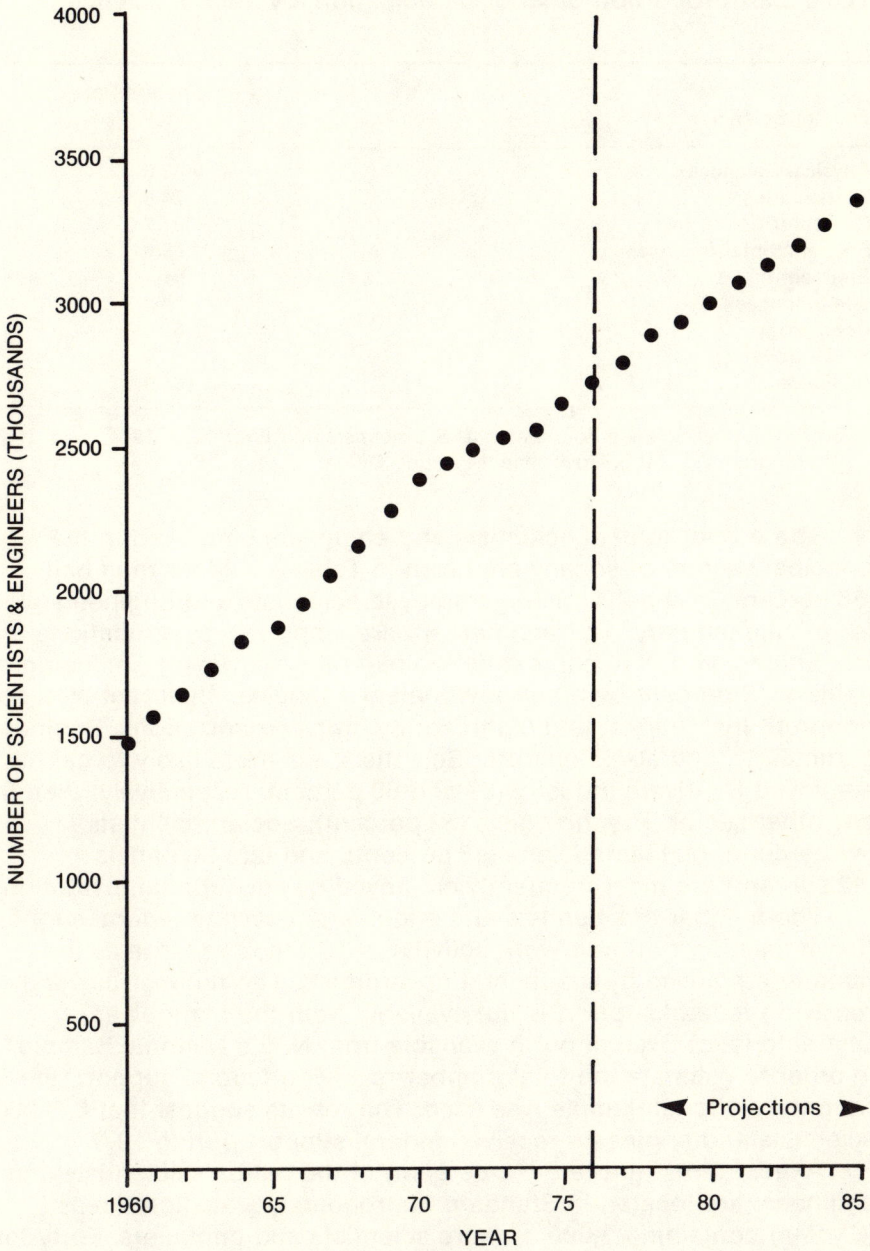

Figure 3.1   Number of U.S. scientists and engineers, 1960–1985.

Table 3.2   Proportion of labor participation, by field of science, 1974.

Field of Science	Participation Rate (%)
Physical Sciences	83.6
Mathematics	74.8
Computer Sciences	97.7
Environmental Sciences	85.6
Engineering	94.1
Life Sciences	70.7
Psychology	65.5
Social Sciences	53.9
All Fields	85.0

Source: National Science Foundation, *U.S. Scientists and Engineers: 1974.* (Washington, D.C.: U.S. Government Printing Office, 1976), p. 19.

The proportions of scientists and engineers employed in the principal sectors of society are given in Table 3.3. More than half (56 percent) of the 2.4 million employed scientists and engineers work for private industry. Thirteen percent are employed by educational institutions, about 10 percent by the federal government (including the military), 9 percent by other government agencies, 4 percent by other nonprofit institutions, and 8 percent by other organizations. Engineers particularly, and also Computer Scientists, are more likely to be employed in private industry (69 and 60 percent, respectively) than in any other sector. Psychologists (52 percent), Social Scientists (47 percent), Mathematicians (46 percent), and Life Scientists (42 percent) are most frequently employed by educational institutions.

The number of scientists and engineers receiving federal support (for at least part of their work activity) is 3.7 times as large as the number employed by the federal government. The number currently receiving federal support is not available from the Manpower Characteristics System but it available from NSF's National Sample.[3] In order to estimate the total number receiving federal support, the ratio from the National Sample was used. The results suggest that 629,000 scientists and engineers received federal support during 1977.

Almost three-quarters (72 percent) of the nation's scientists and engineers are located in Standard Metropolitan Statistical Areas (SMSAs) containing 3,000 or more scientists and engineers. Forty-four percent are in SMSAs containing 20,000 or more scientists and engineers. Only eleven of the fifty states do not contain concentrations of 3,000 or more (see Figure 3.2).

Table 3.3 Proportion of employed scientists and engineers, by sector of employment, 1974 (in percent).

Field of Science	Educational Institution	Federal and Military	Other Government	Nonprofit	Business and Industry	Other
Physical Sciences	23.5	10.0	5.5	5.3	43.4	12.3
Mathematics	46.0	15.2	7.6	2.9	21.0	7.2
Computer Sciences	6.2	7.6	9.2	3.8	60.9	12.4
Environmental Sciences	20.9	17.7	11.3	1.8	39.5	8.8
Engineering	3.1	9.2	9.4	2.2	69.4	6.8
Life Sciences	42.4	19.0	10.4	5.4	16.3	6.6
Psychology	51.8	3.1	8.0	22.5	7.0	7.8
Social Sciences	47.4	11.6	13.6	7.8	14.7	4.9
All Fields	13.3	10.1	9.2	3.8	56.0	7.7

Source: National Science Foundation, *Characteristics of the National Sample of Scientists and Engineers, 1974. Part 2: Employment* (Washington, D.C.: National Science Foundation, 1976), pp. 77–84.

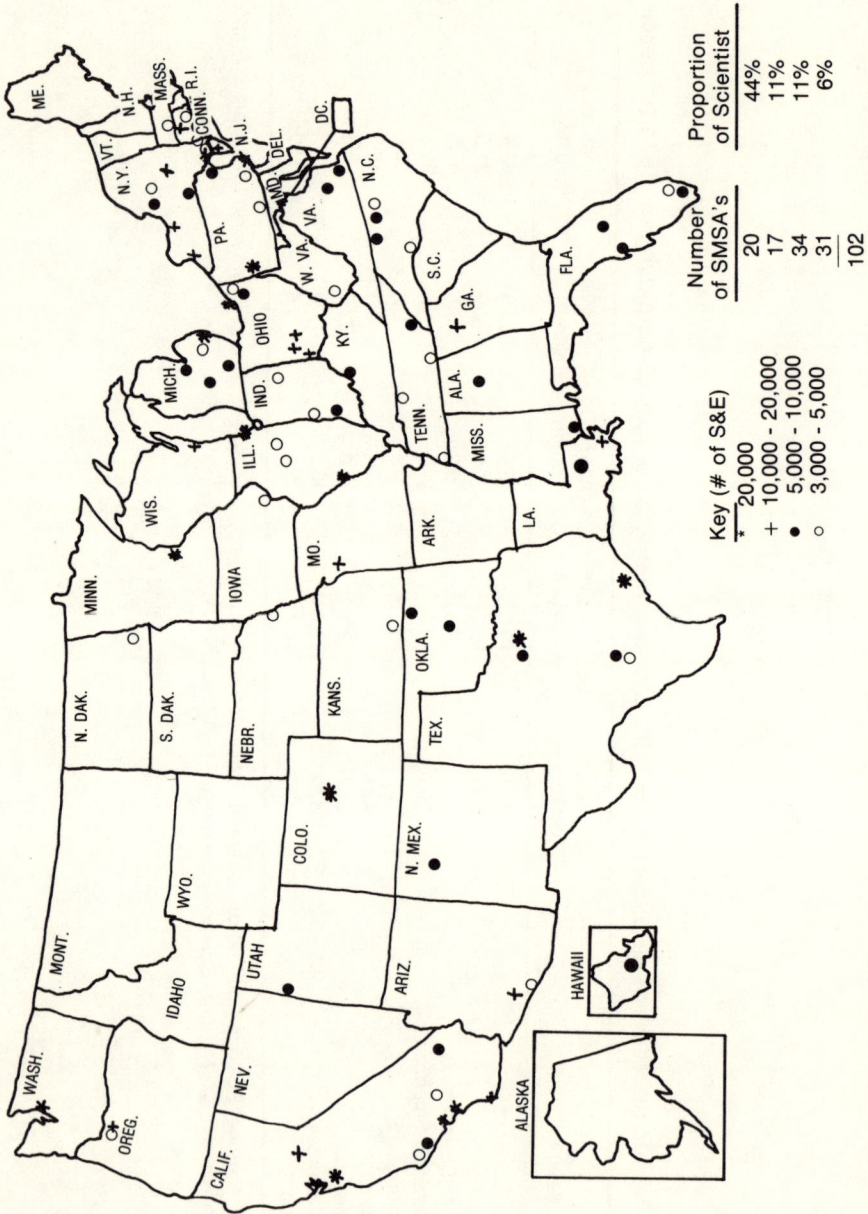

Figure 3.2 Concentration of scientists and engineers in Standard Metropolitan Statistical Areas, 1974.

	Number of SMSA's	Proportion of Scientist
	20	44%
	17	11%
	34	11%
	31	6%
	102	

Key (# of S&E)
* 20,000
+ 10,000 - 20,000
● 5,000 - 10,000
○ 3,000 - 5,000

DESCRIPTION OF JOURNAL AUTHORS

The scientific community is the universe of potential authors of scientific and technical journal articles, but not all scientists are regular or even one-time authors. The characteristics of the author community are somewhat different from the characteristics of the scientific community as a whole.

Price and Gursey, in their *Studies in Scientometrics,*[4] have distinguished among seven groups in the scientific community and have made estimates of their sizes. These groups are:

a. Transients, who publish only during a single year.
b. Recruits, who begin publishing during the year considered and will continue publishing.
c. Terminators, who have published previously but end their publishing during the year.
d. Core continuants, who publish in the year in question and every year over a long period.
e. Noncore publishing continuants, who publish this year and are likely to publish frequently, but not annually, over a long period of time.
f. Nonpublishing continuants, noncore continuants who do not happen to publish in the particular year in question.
g. Nonauthors, who do not publish during the year and are not nonpublishing continuants. This group includes past and future transients who have published or will publish in another year and authors who have terminated previously.

According to Price and Gursey, groups a through e, respectively, make up about 22, 11, 4, 20, and 47 percent of the scientists publishing in a particular year. (Terminators appear to overlap with continuants in this percentage breakdown.) In total, then, 67 percent of the authors in a particular year are continuants, either core or noncore. Nonpublishing continuants are equal in number to about one-third of the publishing authors. It is more difficult to determine the size of the nonauthor group in a given year, although this can be derived by inference from authorship data from several sources.

The most active field of science in terms of authorship is Life Sciences; 17 percent of the scientists published in 1977. Another 6 percent were active but nonpublishing continuants during that year (Table 3.4). The least active fields were Computer Sciences and Engineering, in which only 2.5 percent of the scientists were active authors.

Table 3.4 Authorship, by field of science, 1977.

Field of Science[a]	Number of Articles Published[b] (000)	Average Authors per Article	Number of Article Authorships (000)	Number of Distinct Authors (000)	Number of Scientists or Engineers (000)	Proportion of Authors (%)	Proportion of Nonpublishing Continuants (%)	Total Proportion Active (%)
Physical Sciences	33.7	2.33	78.5	23.8	287	8.3	2.8	11.1
Mathematics	5.8	1.32	7.7	3.2	112	2.3	0.8	3.1
Computer Sciences	6.8	1.49	10.1	3.7	192	1.9	0.6	2.5
Environmental Sciences	13.2	1.55	20.5	7.3	90	8.1	2.7	10.8
Engineering	32.3	2.06	66.5	25.7	1378	1.9	0.6	2.5
Life Sciences	95.0	2.01	191.0	54.6	323	16.9	5.6	22.5
Psychology	11.8	1.74	20.5	5.1	136	3.8	1.3	5.1
Social Sciences	55.1	1.35	74.4	24.6	244	10.1	3.4	13.5
Other Sciences	41.4	1.42	58.8	23.6	N.A.[a]			
All Fields	295.1	1.79	531.2	171.6	2762	6.2	2.1	8.3

Source: King Research, Inc. Author Survey.

[a]All scientists and engineers are allocated to specific fields of science. The proportions for All Fields include articles published in Other Sciences.

[b]In United States only.

Considering all scientists, it can be assumed that from 2 to 17 percent of scientists, depending on the field of science, will author at least one article in a particular year. About 80 percent of these (2 to 14 percent) will be continuants in some sense; they have authored or will author articles in other years. (Included in the 80 percent are Price and Gursey's groups a to e.) Core continuants, or authors who publish every year, account for about 20 percent of the authors in a given year, or about 0.5 to 3 percent of all scientists.

In an additional refinement, Price and Gursey subdivided authors according to the number of articles published. In a small sample, they found that transients produced 1.1 authorships per year and that continuants produced in relation to the length of continuance (for example, two-year continuants produced an average 1.5 articles per year, five-year continuants 3.7 articles per year, and nine-year continuants 4.3 articles). These results correspond well to Price's law, which states that the number of authors with at least n authorships is proportional to $1/(n (k + n))$. (Here k is a parameter of about fifteen authorships per author per lifetime. This marks a boundary between very high and normal production.)

Data have also been collected on the characteristics of authors as compared to the general community of scientists. Comparisons of relevant demographic characteristics are presented in Table 3.5. It shows that authors tend to be more highly educated, are more likely to work in a university environment, and are more likely to be involved in research and/or teaching. The table also displays 1974 demographic data for scientists and engineers. From these data, it appears that the responses to our user survey were somewhat skewed in favor of younger persons with more advanced degrees employed by universities.

Not displayed here are the different demographic characteristics of authors in different fields of science. For example, McDonald found that authors who had published in the journals *Cancer Research* and *Astrophysical Journal* differed significantly in terms of their professional age, (the number of years since they had received their last professional degree).[5] Approximately 58 percent of *Cancer Research* U.S. authors had received their last degree during the previous ten years versus 80 percent of U.S. *Astrophysical Journal* authors.

U.S. authors of journal articles predominantly hold a Ph.D. degree (Table 3.5). Only a small proportion (6 percent) hold another professional degree, such as a medical or veterinary professional degree. In this respect, authors are more highly educated than nonauthors.

Despite pressures to open up U.S. scholarship and research to women, males continue to dominate journal article publishing; only

Table 3.5   Proportion of authors and nonauthors with certain demographic characteristics, 1977 (in percent).

	Authors	Nonauthors[a]	All Scientists, 1974
Highest degree held			
Doctorate	72	37	17
Other professional	6	1	1
Masters	15	24	22
Bachelors	4	31	57
Less than bachelors	3	7	3
Type of employer			
University	62	21	13
Government	15	19	19
Commercial or nonprofit institution	19	43	60
Retired or not employed	1	3	6
Other	3	14	8
Type of work activity[b]			
Research	54	32	30
Teaching	27	20	8
Management	13	30	27
Medical practice	2	1	
Retired or not employed	1	3	6
Other	3	14	34
Age			
Under 30	19	25	10
31–40	45	22	33
41–50	24	21	29
Over 50	13	32	28
Sex			
Male	90	93	95
Female	10	7	5

Source: King Research, Inc. Author Survey; King Research, Inc. User Survey; National Science Foundation, *Characteristics of the National Sample of Scientists and Engineers, 1974. Part 2: Employment* (Washington, D.C.: National Science Foundation, 1976), pp. 00–00.
[a] Only 48 percent of nonauthors responded, so some nonresponse bias could exist.
[b] Individuals are often involved in more than one activity; hence the total exceeds 100 percent.

10 percent of 1977 authors were female. This closely reflects the same pattern for nonauthors, with 7 percent females. Compared with 1974 National Science Foundation figures, however, this may reflect a slight increase in the number of U.S. female scientists and engineers.

Estimates from our journal tracking survey suggest that 57 percent of the articles published in 1977 were based on research or

Table 3.6 Proportion of sources of funding, 1977 (in percent).

Major Funding Sources	Authors	Nonauthors
Government	57	30
Private industry	8	42
Educational institution	27	19
Nonprofit institution	4	3
Other	5	7

Source: King Research, Inc. Author Survey; King Research, Inc. User Survey.

experimentation wholly or partially funded by the federal government. It has also been determined that about 59 percent of the authors of these articles had at least some of their work funded by the federal government. When asked who supplies the largest portion of funds for current projects, authors and nonauthors responded according to the percentages shown in Table 3.6. The largest disparities between authors and nonauthors are in the areas of funding by government (57 percent of authors versus 30 percent of nonauthors) and funding by private industry (8 percent of authors and 42 percent of nonauthors), clear evidence of the strength of the government's contribution to the production of the journal literature.

NUMBER OF JOURNAL ARTICLES PUBLISHED

The total number of articles published in U.S. journals in 1977 was about 382,000. The growth rate in the number of articles has been fairly constant at about 4 percent since 1970, and this trend is expected to continue through 1985. The total number of articles estimated for 1985 is 477,000. The number of these articles published over time is given in Figure 3.3. These statistics represent all articles published in U.S. journals. A distinction must be made between articles published in U.S. journals and those published by U.S. scientists and engineers because a large number of articles (approximately 87,100) that are published in U.S. journals are written by foreign authors, and U.S. authors published approximately 42,200 articles in foreign journals. For the remainder of this section, whenever the number of U.S. (domestic) articles is being discussed, the figures refer to articles written by U.S. authors for U.S. journals. This includes articles with one or more foreign coauthors.

There are several possible implications of the large increases in the amount of scientific and technical literature available to scientists. One possibility is that scientists and engineers are better informed than they

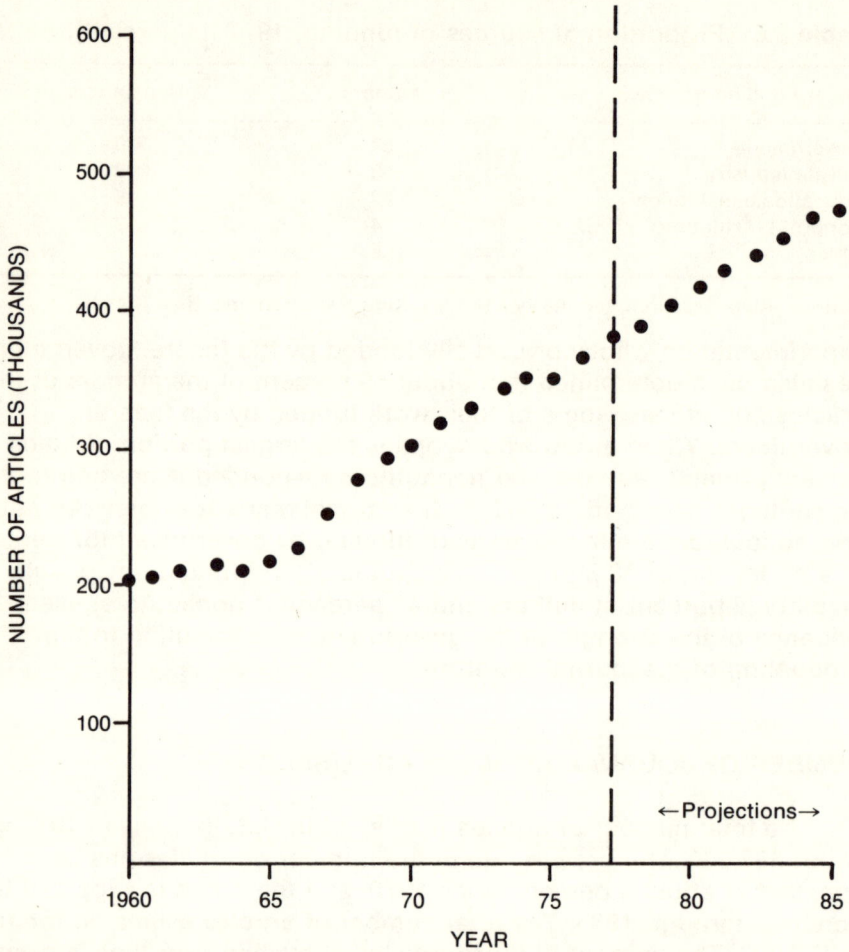

Figure 3.3 Number of U.S. articles published, 1960–1985.

used to be about developments in their own fields. With the great increase in the number of articles published, it may seem that scientists should become increasingly knowledgeable. Another possibility is that scientists are becoming less aware of developments in their fields because they cannot spend the time required to read the vast amounts of periodical literature that are being generated today. The average amount of time scientists spend reading journals appears not to have changed as much since 1960 as the increase in the volume of available literature.

Although there is a large increase in the amount of periodical literature being produced, the number of scientists and engineers is also

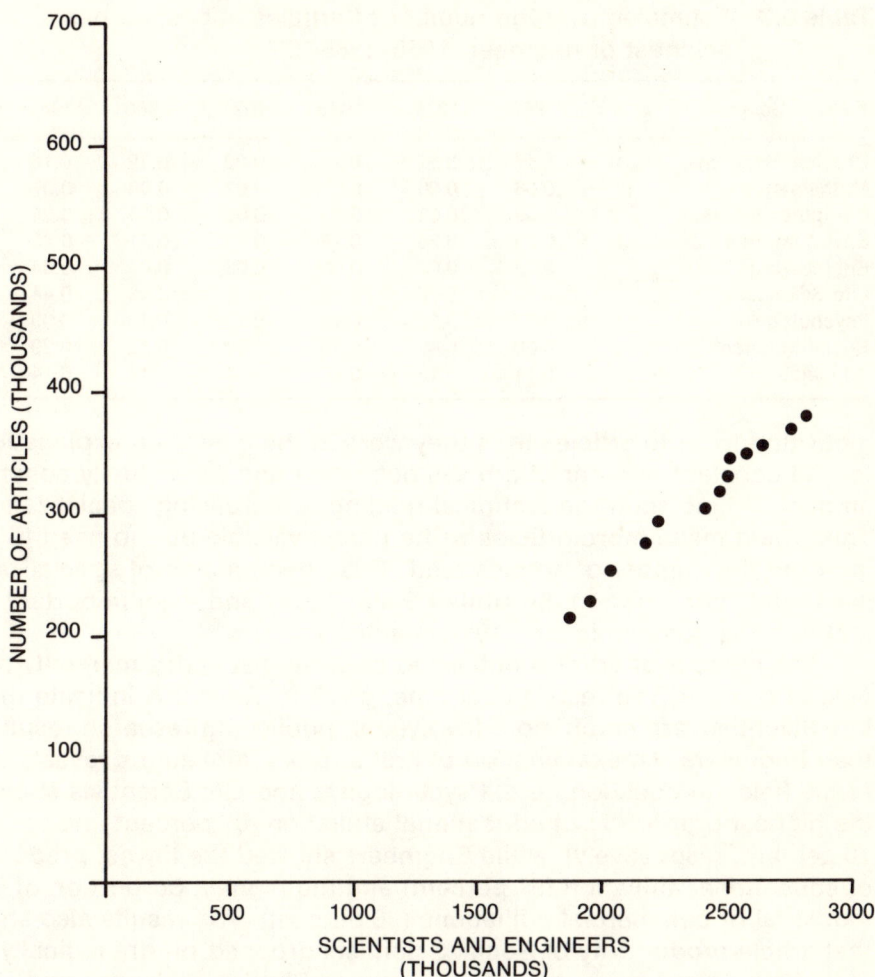

Figure 3.4 Number of journal articles published in the United States as a function of scientists and engineers.

increasing. Between 1965 and 1977 the number of domestic articles increased 76 percent, while the number of U.S. scientists increased 48 percent. This suggests a roughly parallel growth in scientists and engineers and number of articles published. The high correlation of these numbers is shown in Figure 3.4. For the nine fields of science combined, the average number of articles per scientist or engineer changed very little between 1965 and 1977 (for 1965, 0.12 articles per scientist; for 1977, 0.14 articles per scientist). This comparison of articles and scientists shows that perhaps scientists are no more or less

Table 3.7   Estimated average number of articles published per scientist or engineer, 1965–1985.

Field of Science	1965	1970	1975	1977	1980	1985
Physical Sciences	0.17	0.17	0.20	0.20	0.19	0.18
Mathematics	0.08	0.06	0.07	0.07	0.08	0.07
Computer Sciences	0.09	0.03	0.03	0.05	0.04	0.04
Environmental Sciences	0.20	0.25	0.18	0.19	0.21	0.22
Engineering	0.02	0.02	0.03	0.03	0.03	0.04
Life Sciences	0.30	0.30	0.35	0.40	0.42	0.44
Psychology	0.17	0.12	0.10	0.10	0.10	0.09
Social Science	0.60	0.43	0.20	0.22	0.22	0.20
All Fields	0.11	0.12	0.12	0.14	0.14	0.14

motivated to write articles than they were in the past. One explanation for the constant number of articles published and the virtually constant amount of time spent on technical reading is increasing specialization. This would mean more articles written and available but no need to increase the number of articles read. This phenomenon of specialization is certainly occurring in the United States today and is an important factor in the communication system activity.

The number of articles published per scientist varies markedly by field of science. The results over time, given in Table 3.7, indicate that Life Scientists are much more involved in publishing research results than Engineers. An examination of first authors' affiliations reveals that of the fields in question here, Psychologists and Life Scientists showed the highest proportion of educational affiliation (87 percent and 70 percent, respectively), while Engineers showed the lowest proportion of educational affiliation (51 percent) and the highest proportion of industrial or commercial affiliations (26 percent). The results also show that article productivity of Social Scientists dropped off dramatically after 1965. Slight increases are observed in Physical Sciences and decreases in Computer Sciences and Psychology. The last two fields have had significant increases in the number of scientists, which may be a factor.

AUTHORS' SELECTION OF JOURNALS TO WHICH TO SUBMIT ARTICLES

Numerous factors are involved in the process by which scientists select journals to which to submit articles. Uppermost is the selection of a journal with the appropriate subject scope. This and other factors were discussed by Kochen and Tagliacozzo,[6] who stated that authors selected the most prestigious journals with the largest circulation and the shortest publication lag. They also reported research that suggested that in "pre-paradigm" fields authors' selection of an appropriate journal for publishing was more difficult than in fields with more highly developed histories.

In the same article, Kochen and Tagliacozzo described a hypothetical service, primarily for novice authors or authors new to a field. This service would aid in the selection of a journal to which to submit an article. It would make use of a mathematical model that could be used to calculate a score for the desirability or relative preference of a given journal as a publishing outlet for a given article. The model incorporated the following variables: (1) relevance based on readers sharing authors' interests; (2) acceptance rate (authors preferring journals with higher acceptance rates); (3) circulation (the total number of copies perhaps being of secondary importance to the type of readership); (4) prestige, (perhaps measured by journal age); and (5) publication lag (authors preferring a short publication lag). With regard to publication lag (or publication speed, a term preferred by some journal publishers), Garvey, Lin, and Tomita[7] reported that journal article authors who had previously disseminated their research findings via conference and other channels were less interested in speed of article publication than those who had not done so.

A common thread throughout these and other studies is the stated importance of the concepts *prestige, quality, value,* and so forth. These difficult-to-define concepts were among those treated in a study by McDonald[8] investigating the factors, or "attributes," that were related to scientists' selection of particular journals in which to publish their research findings. This study looked at three different types of variables potentially related to authors' decisions:

1. Directly measurable journal characteristics (such as price, frequency of publication, publication speed).
2. Variables describing authors' past experience with the journals (such as number of articles submitted to the journal, number of articles read).
3. Subjectively scored journal "attributes" representing both quantitative and qualitative concepts (such as emphasis on methodology, benefit to career, perceptions of price and publication delay).

Through surveys conducted with authors of *Cancer Research* and *Astrophysical Journal,* it was found that the third group of journal attributes was more effective in discriminating between journals considered for publishing and those not considered for publishing than the first two groups of variables. Based on their perceived importance and on their ability to discriminate among groups of similar journals, the following seven journal attributes were selected for analysis:

> "The proportion of people I wanted to read my paper who were regular readers of this journal"
>
> "The benefit to my career, professional prestige, job status, or salary of publishing in this journal"
>
> "The quality of the printed appearance of papers published by this journal"
>
> "The competence and fairness of this journal's refereeing of submitted papers' scientific content"
>
> "The speed with which this journal publishes a paper once its author is notified of the paper's acceptance"
>
> "The probability of my paper's acceptance by this journal"
>
> "This journal's emphasis on theory"

Author-supplied data regarding the importance of these attributes* was combined with authors' scores† on the same attributes for journals similar to *Cancer Research* and *Astrophysical Journal.* To test the hypothesis that authors' relative preferences for journals can be simulated using basic utility theory concepts, McDonald used the following functional form to calculate utility scores, as suggested by Huber in his review of multiattribute utility theory:[9]

$$U(A_i) = \sum_{n=1}^{N} b_n \, x_{ni}$$

where A_i is the alternative journal under consideration, b_n is a parameter expressing the relative importance weight of journal attribute n, and, x_{ni} is a rating of how much journal i possesses journal attribute n. The quantitiy $U(A_i)$ was viewed as an overall measure of the relative value or utility of an individual journal to an individual scientist. McDonald found that the utility scores calculated using this model were significantly higher for *Cancer Research* and *Astrophysical Journal* than for journals identified by these journals' authors as being similar to them. Additional analysis suggested that the utility scores for other journals considered

---

*Attribute importance was ranked on a zero-to-six scale.
†Supplied by authors on a five-point "very high" to "very low" scale.

for publishing were higher than utility scores for journals not considered for publishing, bearing out the study's initial hypothesis.

This and other types of analysis suggested the following concept areas that might serve as a focus for further study of scientists' communication behavior, especially that related to how authors perceive journals as potential outlets for their research: the subject content or orientation of potential outlets of research; the amount of effort that must be made to satisfy the requests of referees and the editor; the speed with which an article is accepted and published; the control authors have over both the content and the presentation format of their research findings; the identity of the readership the authors want to reach; the benefit to the authors' careers of publishing in a particular journal. As McDonald found, the concern of authors for these areas varies significantly from individual to individual and from field to field. Such variations may be due as much to individual psychological differences as to different communication structures of different fields of science and may merit further research.

JOURNAL AUTHORSHIP ACTIVITIES

Authorship covers the initiating set of activities occurring in the communication cycle, those grouped in our generic functions under composition. In speaking specifically of the development of a scientific or technical journal article under the current system, authorship encompasses literature search, preparing bibliography, writing manuscript, transcribing manuscript, preparing graphics, peer edit and review, revision, proofreading, transcribing revisions, copying the manuscript, dispatching manuscript, ordering reprints, and paying for page charges, alterations, and reprints.

These activities were used as the basis for estimating the cost of authorship in a study conducted for the National Science Foundation.[10] In that study, Life Scientists identified as authors were asked how much time they, their coauthors, and their support staffs spent in preparing a particular journal article. A summary of their responses is shown in Table 3.8, giving a breakdown for the various activities for the initial manuscript and each revision cycle.

These data, combined with knowledge of the numbers of articles and of revision cycles, permit estimation of the total author and support time spent on published articles. In addition, one must also account for rejected articles, including time spent reworking articles to be resubmitted to another journal but excluding time spent on articles rejected and not resubmitted. It is estimated that fifteen hours of a

Table 3.8   Time required for authorship of life science articles, 1974 (in hours)

Activity[a]	Initial Manuscript	First Revision	Second Revision	Third Revision	Fourth Revision	Fifth Revision
Literature search	30.6	4.8	3.5	3.0	2.7	2.5
Preparing bibliography	7.1	1.8	1.7	1.7	1.7	1.7
Writing manuscript	53.5	9.6	5.8	3.6	2.8	2.0
Typing manuscript	7.9	4.8	4.0	3.6	3.5	3.4
Informal internal edit	6.6	3.1	1.9	1.5	1.2	1.0
Formal internal edit	6.9	2.1	2.2	2.2	2.2	2.2
Proofreading typeset manuscript	2.5	1.8	1.7	1.7	1.7	1.7
Subtotal						
Author time	79.1	15.4	10.1	7.5	6.4	5.4
Support time	36.0	12.6	10.7	9.9	9.5	9.2
Proofreading/ alteration of galleys (author)	2.5	2.0	2.4	2.4	2.4	2.4

Source: Lois A. Green and Susan T. Hill, *Editorial Processing Centers: A Study to Determine Economic and Technical Feasibility, Part IV: Survey of Authors, Reviewers and Subscribers to Journals in the Life Sciences.* (Rockville, Maryland: Westat, Inc. and Aspen Systems Corporation, 1974).
[a]Data were not collected for preparing graphics, copying, dispatching, or dealing with reprints.

scientist's time are required, on the average, to rework an article after initial rejection. We assume that the amount of support time spent in preparing articles is approximately the same in other fields as for Life Sciences.

The data in Table 3.8 seem to correspond well to estimates of the total amount of time authors indicate that they (and their coauthors) spent on writing and revising articles. The estimate of this time for all fields of science is 82 hours. The estimates by field of science are given in Table 3.9. The amount of time varies somewhat among fields, with Environmental Sciences highest at 137 hours and Psychology lowest at 66 hours. A study of authors of the *Journal of the National Cancer Institute* reveals that their authors spent an average of 62.3 hours per article and coauthors spent 24.1 hours, for a total of 86.4 hours.[11] This compares closely to the estimated 82 hours found in the Life Sciences.

Table 3.9   Average time spent by authors preparing articles, by field of science, 1977.

Field of Science	Time Spent Preparing Articles (hours)
Physical Sciences	69
Mathematics	122
Computer Sciences	87
Environmental Sciences	137
Engineering	90
Life Sciences	82
Psychology	66
Social Sciences	78
Other Sciences	73
All Fields	82

Source: King Research, Inc. Author Survey.

The estimated acceptance and rejection rates and number of modifications observed in 1977 are given in Table 3.10. The estimated proportion of rejections is 44 percent. The rejection rates vary somewhat among fields of science with Psychology the highest and Environmental Sciences the lowest. The proportion of rejections that are in turn resubmitted to another journal is 28 percent. About 16 percent are not resubmitted but dropped, which shows that all rejected articles are not eventually published elsewhere. For those manuscripts that require revisions, it is estimated that the average number of revisions is 0.9. This number does not vary much among the fields of science.

The flow of manuscripts among authors, publishers, editors, and reviewers is time-consuming and complex. The average number of months that elapse from the time an article is first submitted for publication to the time it is published is 10.3 months. This number varies substantially among journals as well as among the fields of science, with Mathematics the highest at 20.5 and Other Sciences the lowest at 5.8 months. A mathematical model used to describe the flow of manuscripts among the participants (authors, publishers and reviewers) is presented in the next section.

A MODEL OF MANUSCRIPT FLOW

After an author has submitted a scientific or technical manuscript to a particular journal, he or she is likely to have a number of additional

Table 3.10 Rejection rates, number of revision cycles, and publication delay, by field of science, 1977.

Field of Science	Proportion of Articles Rejected (%)	Proportion Submitted to Another Journal (%)	Proportion Not Resubmitted (%)	Average No. of Revision Cycles	Average Publication Delay (months)
Physical Sciences	19	10	9	0.7	8.0
Mathematics	35	22	13	0.6	20.5
Computer Sciences	38	24	14	0.9	10.6
Environmental Sciences	19	11	8	1.2	14.4
Engineering	34	17	17	0.8	9.0
Life Sciences	48	33	15	0.9	12.1
Psychology	71	44	27	1.1	12.1
Social Sciences	59	36	23	1.0	10.3
Other Sciences	67	45	22	0.6	5.8
All Fields	44	28	16	0.9	10.3

Source: King Research, Inc. Author Survey.

interactions with the publisher, many of which involve the actual physical transfer of the manuscript. At any of these stages, there is the possibility that the process will be terminated and, thus, that the article will not be published by that journal. The publisher may, before the manuscript is reviewed, return it to the author as unacceptable because of subject matter or perhaps format. Once the manuscript has passed this preliminary stage, it goes on to a subject editor and then to reviewers, who consider the technical content of the paper in some detail. The subject editor and/or reviewers may accept, reject, or recommend modifications to the article. If the article is rejected or modifications recommended, the author may submit the article to another journal. If the author chooses to make the suggested modifications, there may be several cycles of interaction between author and publisher before the manuscript is unconditionally accepted.

Once a manuscript has been accepted, additional interactions will take place. Proofs, usually galley and page proofs, will be sent to the author for review and returned with any corrections marked. When the article is published, reprint copies are often sent to the author. These may be sent without charge, provided in return for payment of page charges, charged for directly, or a combination of these methods. When the author is charged for page charges, reprints, or author alteration, billing and payment become additional author-publisher interactions.

In order to describe more fully the passage of a manuscript from the author into a published form, a model was developed as part of the Westat and Aspen study of Editorial Processing Centers (EPC Study).[12] This model is stochastic, in the sense that one cannot predict with certainty the length of time required to perform any of the article preparation functions, or whether the article will be rejected, accepted, or modified. However, all of the operations of the system can be described by probability distributions so that one can, nevertheless, treat with considerable accuracy the behavior of the system under a large input of articles. The proportion of manuscripts that flow in different directions may be thought of as probabilities.

It is convenient to think of the system as composed of stations and activities, which are performed at the stations. The stations represent locations at which functions related to an article take place. A copy of a manuscript does not always pass from station to station, but the responsibility for acceptance, rejection, or modification may.

Consider a simplified model for a scientific journal consisting only of stations representing the author, managing editor, subject editor, technical reviewer, rejection, publication in the original journal, and submission to another journal. We will arbitrarily set two revisions as the

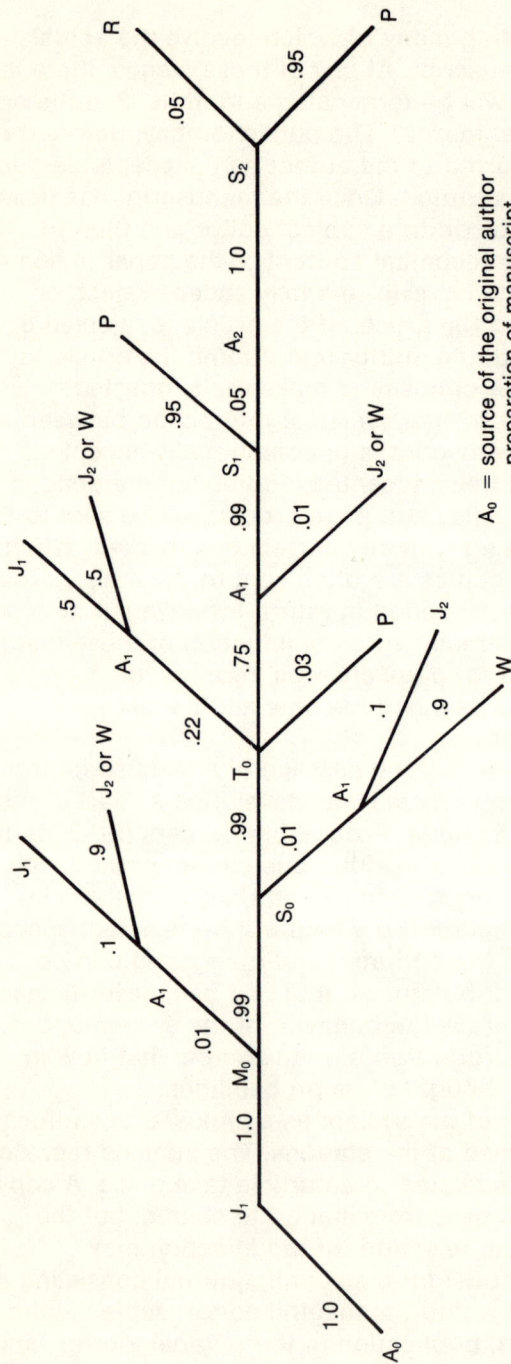

Figure 3.5 Tree diagram of transition probabilities.

A_0 = source of the original author
preparation of manuscript
A_1 = return to the author,
J_1 = decision to submit to the first journal,
J_2 = decision to submit to a second journal,
M_0 = original managing editor review,
M_1 = subsequent reviews by the managing editor
P = publish,
R = reject and drop,
S_0 = original subject editor review,
S_1 = subsequent review by the subject editor,
T_0 = original technical review,
T_1 = subsequent technical reviews,
W = withdraw,

maximum number of times a manuscript will be resubmitted, and require that the manuscript either be rejected, go to another journal, or be published at that point. More than two revision do take place occasionally but not frequently. Since, in general, the probability of passage from the managing editor to each of the other possible stations is dependent on the number of times the article is revised,* we have put in a separate station for each time it is submitted for review and for each time an article is sent back to the author. We have thus constructed a society where the managing editor receives the manuscript once; where the manuscript is sent to a reviewer once; and, where the author revises the manuscript no more than twice. A tree of the transition probabilities for this model is shown in Figure 3.5. Some examples of things that can be determined from the transition probabilities are: What proportion of the original articles will end up in each of the absorbing states, publish (P) and reject (R)? How many times, on the average, must the editor take action on the article?

The probability of a manuscript submitted to the first journal being published in that journal (state P) is:

$$
\begin{aligned}
P\ (P) =\ & [(.5) \times (1.0) \times (.99) \times (.99) \times (.75) \times (.99) \times (.03)] + \\
& [(.5) \times (1.0) \times (.99) \times (.99) \times (.75) \times (.99) \times (.95)] + \\
& [(.5) \times (1.0) \times (.99) \times (.99) \times (.75) \times (.99) \times (.05) \times \\
& (1.0) \times (.95)] = 0.38
\end{aligned}
$$

The probability of a manuscript being rejected completely (state R) is:

$$
\begin{aligned}
P\ (R) =\ & (.5) \times (1.0) \times (.99) \times (.99) \times (.75) \times (.99) \times (.05) \times \\
& (1.0) \times (.05) = 0.001
\end{aligned}
$$

The probability of the manuscript being withdrawn by author and either submitted to a second journal (state J_2) or dropped (state W) is:

$$
\begin{aligned}
P(J_2) =\ & .5 + [(.5) \times (1.0) \times (.01) \times (.9)] + [(.5) \times (1.0) \times (.99) \times \\
& (.01) \times (.9)] + [(.5) \times (1.0) \times (.99) \times (.99) \times (.22) \times \\
& (.5)] + [(.5) \times (1.0) \times (.99) \times (.75) \times (.01)] = 0.57
\end{aligned}
$$

The average number of times that the subject editor sees the article can be determined from the probabilities that he or she sees it once, twice, or three times.

Such a model can be computerized and used to simulate a range of input, process, and output conditions. It can also be adapted to include

*Thus, the model does not exhibit the independence property of a Markov chain.

considerations of timing and of querying by participants of one another; thus, its use can be extended to answer a wide variety of questions. The critical requirement for use of the model is appropriate data on at least the transition probabilities and, additionally, on timing and queries.

Some limited data are available on the various probabilities associated with the processing of a manuscript submitted to a journal. The two major categories of data are acceptance/rejection rates and the number of revision cycles.

The decision to reject, accept, or accept with revisions can take place at any of several stages, as indicated in the model. In the EPC Study, seven society publishers of twenty Life Science journals were asked such questions as:

What percentage of incoming manuscripts are immediately returned to the author by the managing editor for reasons of style or form?	(>1 to <5%)
After acceptance of a manuscript by an editor, what percentage (if any) of the manuscripts will the managing editor return to the author for reasons of style or form?	(2 to 95%)
What percentage of all incoming manuscripts are rejected?	(7 to 62%, median 45%)
What percentage of the articles returned to the author for modification are later withdrawn (either by the author or by default)?	(1 to <10%)

The responses received are shown in parentheses after the questions. They vary greatly and were combined (as a weighted average) for the purposes of cost modeling in the EPC Study. Modifications to the values shown on the tree diagram of Figure 3.5, made for the current study, were based on additional information developed on rejection rates from the author survey.

SPECIAL CONSIDERATIONS IN THE AUTHOR AND PUBLISHER INTERFACE

In addition to writing their manuscripts, authors may, in some instances, be asked by a publisher to pay page charges, reprint charges, or alteration fees. Alteration fees are generally imposed in special instances such as when an author requests changes after page proofs have been created. These charges are generally used more as a control mechanism than as a cost-recovery device, and even in the aggregate do not involve significant amounts of money. Page charges and reprint fees, however, can make up a significant percentage of a publisher's income.

The term *page charge* refers to "the practice of requesting from the institution supporting the research reported in a published article a payment of a certain number of dollars per published page of the article".[13] The page charge reflects, in a sense, the editorial costs associated with preparation of the manuscript for reproduction and distribution. Page charges are usually voluntary rather than obligatory, and they are levied by society and other nonprofit journals.

In our survey, authors were asked how much they or their employers paid in page charges or special fees to help support journal publication. The responses for 1977 articles are shown in Table 3.11. The page charges dominate the volume of fees although reprint/preprint fees are fairly common as well. Such fees are far more common in the fields of Engineering, Environmental Sciences, and Physical Sciences and are rare in the Social Sciences and Psychology. As indicated, when all articles and all fields of science are considered, the average page charge is $113 per article. Only a third of all authors pay page charges, however, so that the average actual payment is $208. This figure has remained fairly constant, in current dollars, over the years 1968 to 1977.[14] Among the fields of science, the percentage of authors reporting payment of one kind or another ranged from 24 percent in the Social Sciences to nearly 100 percent in Environmental Sciences. The average payment made by or on behalf of an author paying page charges ranged from $38 in Social Sciences to $547 in Engineering.

Only 18 percent of the authors who paid page charges paid them out of their own pockets. Of those authors who paid page charges or other fees, it was found that most were actually paid by the federal government as employer or through a grant or contract (68%), with other employers (33%) second. This picture did not differ appreciably among the fields of science.

In his survey of authors of the U.S. society-sponsored journals *Cancer Research* and *Astrophysical Journal,* McDonald[15] found that the assessment of page charges was not an important factor in authors' selection of a journal to which to submit an article. This apparently reflected the very low proportion of U.S. authors who pay page charges with their own funds. According to an Institute of Physics study,[16] however, page-charge assessment may be a more important consideration for non-U.S. authors due to the lower availability of research funding in other countries.

Table 3.11 Proportion of articles involving author fees, by field of science, 1977.

Field of Science	Proportion of Articles				Average Fees per Article All Authors	Average Fees per Article Paying Authors
	Submission Fees (%)	Page Charges (%)	Reprint/ Preprint Fees (%)	Revision Fees and Other (%)		
Physical Sciences	9	76	34	6	$169	$274
Mathematics	1	32	7		40	87
Computer Sciences	2	14	7	3	40	127
Environmental Sciences	1	91	65	9	142	242
Engineering	2	84	40		301	547
Life Sciences	2	35	22	2	83	167
Psychology	7	24	14	8	81	108
Social Sciences		24		2	10	38
Other Sciences	10	6	4	3	20	61
All Fields	4	38	20	3	113	208

Source: King Research, Inc. Author Survey.

Table 3.12 Average number of reprints distributed by authors, by field of science, 1975.

Field of Science	Average No. of Reprints Distributed
Physical Sciences	51
Mathematics	21
Computer Sciences	31
Environmental Sciences	92
Engineering	39
Life Sciences	110
Psychology	65
Social Sciences	30
Other Sciences	45
All Fields	69

Source: King Research, Inc. Author Survey.

Authors may also receive reprints of their articles from the publisher, which they in turn can distribute to colleagues and others. They may be charged for reprints or receive at least an initial amount free or in return for payment of page charges. These reprints, and others which are obtained by interested scientists directly from the publisher, are an important source of journal information and are highly used in relation to the number of reprint copies printed.

The average number of reprints received by an author in 1975 was estimated at 142 in the Statistical Indicators study.[17] Of these, less than half were distributed (within the first year after publication). Estimates of distribution by authors for the different fields of science are shown in Table 3.12.

Many of the articles published in the literature report federally funded research. Over all fields, the proportion of such articles is about 52 percent* of all scientific and technical articles. The number of articles reporting federally funded research is given in Table 3.13. A high proportion of articles reporting on federally funded research are found in the Physical Sciences, Environmental Sciences, and Engineering. All of these fields also frequently require page charges.

COST OF JOURNAL AUTHORSHIP

The costs involved in journal authorship are primarily those of time spent by the author or coauthors and by support staff. In some

*Note that the proportion given in Table 3.6 (57%) included partially funded research as well as articles published outside the United States.

Table 3.13 Number of articles written that report federally funded research, by field of science, 1977.

Field of Science	Articles Based on Federally Funded Research	Proportion of All Articles (%)
Physical Sciences	26,467	78.6
Mathematics	2,613	44.5
Computer Sciences	2,938	43.3
Environmental Sciences	9,723	73.6
Engineering	22,443	69.3
Life Sciences	53,680	56.5
Psychology	5,635	47.8
Social Sciences	11,179	20.5
Other Sciences	19,434	46.9
All Fields	154,112	52.3

Source: King Research, Inc. Author Survey.

instances, page charges or other fees are also paid by the author and these are considered an authorship cost.

The salaries of scientists and support personnel are, of course, significant in determining the cost of authorship. The salaries of scientists are of particular importance, since, as will be seen later, costs associated with their time heavily dominate the total cost of communication. While the makeup of the author community differs somewhat from that of the general scientific community, median salaries are assumed to be the same for each group in this study.

Median salaries for five-year intervals between 1960 and 1985 for the nine fields of science of interest and for scientists overall are given in Table 3.14. The median salary does not vary widely among fields. To convert the salaries given in the table to an hourly basis, it was assumed that scientists spend somewhat more than a normal work week in the performance of all their duties, including authorship and research. This is because activities such as authorship and researching are often extended to evenings and weekends and thus extend the scientist's total time commitment. The estimate of average total time spent is 48 hours per week, or 2,496 hours per year.

The time estimates derived earlier for the various authorship activities, time for reworking rejected articles for resubmission to a different journal, and data describing manuscript flow lead to the estimates of authorship costs presented in Table 3.15. The average and

Table 3.14   Median salaries for scientists and engineers, by field of science, 1960–1985 (in thousands).

Field of Science	1960	1965	1970	1975	1977	1980	1985
Physical Sciences	$9.9	$11.7	$15.5	$21.1	$24.2	$28.9	$36.4
Mathematics	9.0	11.6	14.6	21.4	24.6	29.3	37.2
Computer Sciences	8.8	11.8	16.5	19.9	22.9	27.2	34.5
Environmental Sciences	8.9	10.9	14.9	21.7	25.0	29.7	37.7
Engineering	9.6	11.3	15.2	21.0	24.1	28.7	36.6
Life Sciences	8.8	10.9	14.5	19.2	22.1	26.3	33.5
Psychology	8.0	10.9	15.0	21.0	24.1	28.7	36.4
Social Sciences	9.5	11.9	14.5	21.8	25.1	29.9	38.0
All Fields	9.4	11.4	15.1	20.8	24.0	28.6	36.3

Table 3.15   Authorship costs by field of science, 1977.

Field of Science	Average Cost per Article	No. of Articles by U.S. Authors (000)	Total Cost (millions)
Physical Sciences	$2,372	33.7	$ 79.9
Mathematics	2,447	5.8	14.3
Computer Sciences	2,190	6.8	14.9
Environmental Sciences	2,523	13.2	33.3
Engineering	2,394	32.3	77.5
Life Sciences	2,098	95.0	199.4
Psychology	2,381	11.8	28.1
Social Sciences	2,563	55.1	139.9
Other Sciences	2,561	41.4	106.2
All Fields	2,337	295.1	688.9[a]

Source: King Research, Inc. Author Survey.
[a]Not equal due to rounding.

total costs of authorship are both very high. These costs are often ignored in discussions of publishing costs because they are largely hidden or donated.

A smaller but still significant element in total expenditures is that of page charges and other fees. Table 3.11 indicated the proportion of domestic articles involving author fees and the per article cost. These are combined in Table 3.16 to show the total expenditures—about $35 million over all fields.

Table 3.16   Total expenditures for author fees, by field of science, 1977.

Field of Science	Number of Articles Involving Authors Fees (000)	Average Fee per Article	Total Expenditures (millions)
Physical Sciences	27.0	$274	$ 7.4
Mathematics	3.5	87	0.3
Computer Sciences	2.3	127	0.3
Environmental Sciences	10.7	242	2.6
Engineering	23.1	547	12.7
Life Sciences	58.4	167	9.8
Psychology	9.6	108	1.0
Social Sciences	15.5	38	0.6
Other Sciences	14.5	61	0.9
All Fields	164.6	208	35.5

CONTRIBUTION OF FEDERAL FUNDING TO AUTHORSHIP

Table 3.13 indicated the number of articles written based on research or experimentation at least partially funded by the federal government. It is assumed that this is the number of articles authored under federal support. The authorship cost per article is indicated in Table 3.15, and the two data elements are combined in Table 3.17 to indicate federally supported authorship costs.

The federal government is also the source of page charges paid to support publication of many articles. Of the articles involving page charges (see Table 3.11), 67 percent involved the federal government as a source of funds. The total dollar expenditure of the federal government, by field, is also reported in Table 3.17.

REFERENCES

1. Estimate based on the National Science Foundation's Manpower Characteristics System.
2. National Science Foundation, *U.S. Scientists and Engineers: 1974* (Washington, D.C.: Government Printing Office, 1976).
3. Estimate based on the National Science Foundation's 1976 National Sample.

Table 3.17 Federally supported authorship costs, by field of science, 1977.

Field of Science	Costs of Author and Support Time			Authors Fees		
	Number of Articles Involved (000)	Cost Per Article	Total (millions)	Number of Articles Involved (000)	Expenditure per Article	Total (000)
Physical Sciences	26.5	2,372	$ 62.8	23.7	$274	6,500
Mathematics	2.6	2,447	6.4	0.9	87	80
Computer Sciences	2.9	2,190	6.4	0.3	127	40
Environmental Sciences	9.7	2,523	24.5	8.6	242	2,100
Engineering	22.4	2,394	53.7	17.6	547	9,600
Life Sciences	53.7	2,098	112.6	25.2	167	4,200
Psychology	5.6	2,381	13.4	0.8	108	90
Social Sciences	11.2	2,563	28.7	2.6	38	100
Other Sciences	19.4	2,561	49.8	1.5	61	90
All Fields	154.1	2,337	358.3	81.2	208	22,800

4. D. S. Price and S. Gursey, *Studies in Scientometrics. I. Transience and Continuance in Scientific Authorship* (New Haven, Connecticut: Yale University, 1974).
5. D. D. McDonald, *Interactions Between Scientists and the Journal Publishing Process* (Rockville, Maryland: King Research, 1979).
6. M. Kochen and R. Tagliacozzo, "Matching Authors and Readers of Scientific Papers." *Information Storage and Retrieval* 10:197–210 (1974).
7. W. D. Garvey, N. Lin, and K. Tomita, "Research Studies in Patterns of Scientific Communication. III. Information Exchange Processes Associated with the Production of Journal Articles." *Information Storage and Retrieval* 8:207–221 (1972).
8. McDonald, *Interactions Between Scientists.*
9. G. P. Huber, "Methods for Quantifying Subjective Probabilities and Multi-Attribute Utilities." *Decision Sciences* 5:430–458 (1974).
10. L. A. Green and S. T. Hill, *Editorial Processing Centers: A Study to Determine Economic and Technical Feasibility. IV. Survey of Authors, Reviewers and Subscribers to Journals in the Life Sciences* (Rockville, Maryland: Westat and Aspen Systems Corporation, 1974).
11. D. W. King, D. D. McDonald, and C. H. Olsen, *A Survey of Readers, Subscribers, and Authors of the Journal of the National Cancer Institute* (Rockville, Maryland: King Research, 1978).
12. L. H. Berul, D. W. King, and J. G. Yates, *Editorial Processing Centers: A Study to Determine Economic and Technical Feasibility* (Rockville, Maryland: Westat and Aspen Systems Corporation, 1974).
13. Capital Systems Group, *Page Charge Policies and Practices in Scientific and Technical Publishing: A Historical Summary and Annotated Bibliography* (Rockville, Maryland: Capital Systems Group, 1976).
14. D. W. King, D. D. McDonald, N. K. Roderer, C. G. Schell, C. G. Schueller, and B. L. Wood, *Statistical Indicators of Scientific and Technical Communication (1960–1980),* 2d ed. (Rockville, Maryland: King Research, 1977).
15. McDonald, *Interactions Between Scientists.*
16. Institute of Physics, *Author/Subscribers Survey: Summary of Results* (Bristol, England: The Institute, 1976).
17. King, McDonald, Roderer, Schell, Schueller, and Wood, *Statistical Indicators.*

Chapter 4

Publishing Scientific and Technical Journals

SIZE AND GROWTH OF JOURNAL PUBLISHING

A distinction is made between journals and other periodicals in this chapter. Scientific and technical journals are primary formal vehicles for communicating information and research results. They often employ a peer review or refereeing process to aid in screening manuscripts and for editorial quality control. They also appear in major bibliographic publications and data bases. While their focus is on reporting research findings, methods, concepts, or other scholarly information, they often publish other forms of professional communication such as letters, book reviews, editorials, news, and advertising. Other periodicals (nonjournals) primarily publish less scholarly information such as trade information or news. These other periodicals include bulletins, newsletters, and industrial trade periodicals.* In the remainder of this chapter we will discuss aspects of the larger periodical industry as well as the journal industry. Our detailed models, discussed in later chapters, are concerned only with the journal subsets as the major form for communicating research results.

Estimates of the number of journals and other periodicals published in the United States were made by combining lists of publications[2,3] and complementing these lists by sampling from *Ulrich's International Periodicals Directory.*[4] By this procedure, we estimate that 4,447 scientific and technical journals and 4,468 other scientific and technical periodicals were published in the United States in 1977. This compares to an estimated 54,700 scientific and technical periodicals published worldwide based on British Library Lending Division data.

The growth of the number of U.S. journals and periodicals has been steady, whereas worldwide growth appears to be more dynamic.† The

*A more complete definition of scientific and technical journals and other periodicals is given in King et al.[1]

†Some caution should be maintained concerning worldwide estimates. First, there are no consistent sources of data. Second, the best recent source has been the British Library Lending Division and their data may reflect improved coverage as well as true growth.

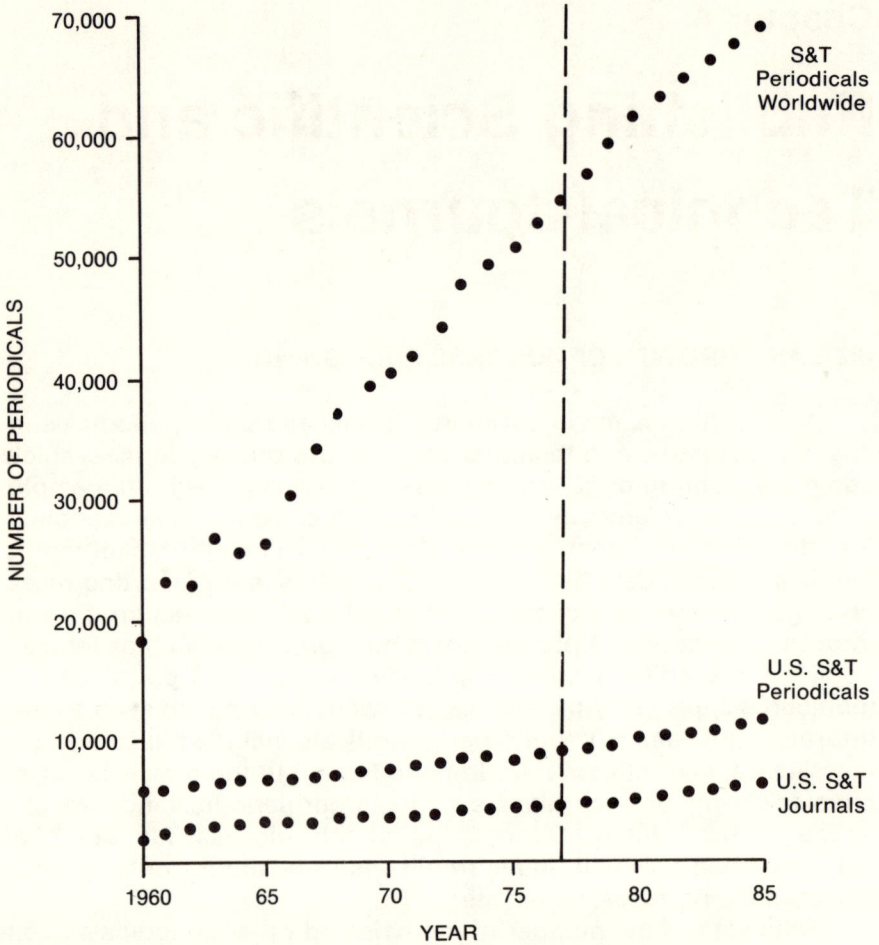

Figure 4.1 Growth of worldwide scientific and technical periodicals, U.S. scientific and technical periodicals, and U.S. scientific and technical journals, 1960–1985.

estimated growth of periodicals and journals is shown in Figure 4.1 with three growth curves: the number of scientific and technical periodicals worldwide, the number of scientific and technical periodicals in the United States, and the number of scientific and technical journals in the United States.

The historical growth of scientific and technical periodicals is not known exactly. Data are available on the birth of new journals, but good data have not been kept for periodicals that have died over the years. Price,[5] writing in 1961, stated that the number of periodicals appears to double about every fifteen years and that this rate had been remarkably constant over the centuries. This rate would average about 5 percent annually. Other sources of data on growth of periodicals support ranges of about 2 to 12 percent per year. In 1970, a task group of the National Academy of Sciences[6] estimated the current birth rate of new journals to be about 2.5 percent a year, although this was considered to be an underestimate. They found mortality to be negligible. The International Council of Scientific Unions (ICSU)[7] found the average annual rate of growth for Physical Sciences journals to be 12 percent over the years 1910 to 1960 and 5.5 percent in the Biological Sciences.[8] Our estimates for the period 1960 to 1975, based largely on British Library Lending Division data, yield an average of 2.7 percent annually. Thus, these estimates are in the same order of magnitude of those observed by others.

In the United States, there is some evidence of considerable growth in number of journals historically. A picture of this growth up to 1959 was presented by the National Science Foundation,[9] and is repeated in Figure 4.1. Forecasts have been made from a time series equation:

$$Y = (3.19 - 2.06X + 0.172X^3 - 0.002X^4) \times 10$$

where Y is the estimated number of U.S. S&T journals and X designated the year ($X = 1$ for 1849, 2 for 1958, 3 for 1869, . . .).

Growth experienced earlier appears to be dampening somewhat since the average annual increase over ten years at the turn of the century was 6.7 percent and the comparable figure for the middle of the century was 2.5 percent. Our estimates of the average annual growth from 1960 to 1977 are 2.7 percent for scientific and technical journals and 1.2 percent for other scientific and technical periodicals.

The total number of periodicals varies substantially among the fields of science (Table 4.1). Social Sciences and Life Sciences are estimated to produce the most periodicals (3,321 and 2,149, respectively), while Mathematics and the Computer Sciences produce only 121 and 134 periodicals, respectively. Later in this chapter we will show that the number of articles published by field is highly correlated over the years with the number of scientists and engineers, although the number of scientists per journal varies significantly from field to field. The proportions of journals and of other periodicals also vary markedly. The Engineering, Environmental Sciences and Social Sciences periodicals have a relatively low proportion of journals (36, 39, and 43%, respectively). The fields of science with a particularly high proportion of journals are Physical Sciences (67%), Mathematics (88%) and Psychology (84%), perhaps reflecting more formal communication

Table 4.1 Number of U.S. S&T periodicals, journals and other periodicals, by field of science, 1977.

Field of Science	All S&T Periodicals	Journals	Other Periodicals
Physical Sciences	415	278	137
Mathematics	121	106	15
Computer Sciences	134	68	66
Environmental Sciences	600	234	366
Engineering	1,297	417	826
Life Sciences	2,149	1,318	831
Psychology	270	228	42
Social Sciences	3,321	1,428	1,893
Other Sciences	608	316	292
All Fields	8,915	4,447	4,468

Source: King Research, Inc. Journal Tracking Survey.
Note: The term *Field of Science* and each of the specific designations are adopted as quasiproper names from "NSF Fields of Science Defined."

patterns (for example, more common use of article refereeing) in these fields.

We have projected the numbers of journals published by field of science up to 1985.* One way to describe the journal publishing activity is by the ratio of number of scientists or engineers per journal. Estimates of this number are given in Table 4.2 for the years 1965, 1977, and 1985. In 1977 the most active fields of science were the Environmental, Life, and Social Sciences where the number of scientists per journal is less than 500. At the other extreme, Computer Sciences and Engineering have nearly 3,000 persons per journal. This suggests that the relative importance of journals in these fields varies substantially. Part of this is due to type of employment and work done. The fields with heavy concentrations in academia have greater activity in journal publishing, that is, more journals per scientist or engineer.

Because much of our analysis emphasizes the role of scientific and technical journals in the communication system, Table 4.3 presents data on the number of these journals and the total number of articles included in them for each of the years between 1960 and 1985. Total number of articles in 1977 is about 382,000, or about 86 articles per journal. The numbers of articles per journal vary somewhat among the fields, ranging from 40 in Social Sciences to 207 in Physical Sciences in 1977. Over the years, there have been gradual increases and decreases

*The projections are based on a linear regression time series model with independent variables including number of scientists or engineers and autocorrelated previous years values.

Table 4.2 Number of scientists and engineers per journal, by field of science, 1965, 1977, and 1985.

Field of Science	Year	Number of Scientists and Engineers (000)	Number of Journals	Number of S&E's per Journal
Physical Sciences (1)	1965	253	191	1,325
	1977	287	278	1,032
	1985	335	294	1,139
Mathematics (2)	1965	52	64	813
	1977	112	106	1,057
	1985	136	138	985
Computer Sciences (3)	1965	31	36	861
	1977	192	68	2,824
	1985	284	92	3,087
Environmental Sciences (4)	1965	50	182	275
	1977	90	234	385
	1985	120	259	463
Engineering (5)	1965	1,182	412	2,869
	1977	1,378	471	2,926
	1985	1,578	516	3,058
Life Sciences (6)	1965	212	856	248
	1977	323	1,318	245
	1985	388	1,648	235
Psychology (7)	1965	37	114	325
	1977	136	228	596
	1985	212	336	650
Social Sciences (8)	1965	50	900	55
	1977	244	1,428	171
	1985	296	1,678	176
Other Sciences (9)				
All Fields (10)	1965	1,867	3,010	620
	1977	2,762	4,447	621
	1985	3,349	5,304	631

among the fields, with the new overall effect a general increase in the number of articles per journal.

One very important characteristic of scientific and technical journals is their number of subscriptions. We have subdivided journals into those that have less than 3,000 subscriptions (small), between 3,000 and 10,000 subscriptions (medium), and over 10,000 subscriptions (large). The numbers of journals falling into these categories are 2,271, 1,580, and 596, respectively. The growth of number of journals in these three categories is given in Figure 4.2. The growth in large journals is much more rapid than in medium and small journals. In fact, the average number of subscriptions to medium journals appears to be dropping somewhat.

Table 4.3 Number of S&T journals, journal articles, and articles per journal, by field of science, 1960–1985.

		1960	1961	1962	1963	1964	1965	1966	1967	1968	1969	1970	1971	1972	1973	1974	1975
(1)	Journals	132	140	150	162	177	191	204	216	229	242	249	254	259	265	268	271
	Articles	29,088	30,436	33,438	36,362	40,022	43,161	46,125	48,828	50,829	51,602	52,033	54,556	56,723	56,532	55,360	55,996
	Article/Journals	220	217	223	224	226	226	226	226	222	213	209	215	219	213	207	207
(2)	Journals	53	53	54	57	60	64	67	72	76	80	83	86	89	92	95	98
	Articles	2,756	2,809	2,970	3,306	3,600	4,224	4,690	5,400	6,080	6,880	7,387	7,654	7,654	7,636	7,790	8,134
	Article/Journals	52	53	55	58	60	66	70	75	80	86	89	89	86	83	82	83
(3)	Journals	24	25	27	29	32	36	39	42	45	48	50	51	55	58	60	62
	Articles	2,784	2,850	3,078	3,451	3,872	4,464	4,719	4,956	5,310	5,904	6,400	6,885	7,590	7,830	7,920	8,184
	Article/Journals	116	114	114	119	121	124	121	118	118	123	128	135	138	135	132	132
(4)	Journals	156	156	159	166	173	182	192	200	203	205	210	214	218	223	228	228
	Articles	7,284	7,564	7,923	8,746	9,436	9,826	11,248	14,614	17,245	17,145	15,558	14,516	14,420	14,350	14,262	14,732
	Article/Journals	47	48	50	53	55	54	59	73	85	84	74	68	66	64	63	65
(5)	Journals	362	369	375	385	399	412	424	432	439	450	460	469	469	462	454	447
	Articles	28,236	28,782	28,125	28,105	28,329	30,076	32,648	34,128	35,559	36,900	39,100	42,210	44,086	44,352	44,492	43,806
	Article/Journals	78	78	75	73	71	73	77	79	81	82	85	90	94	96	98	98
(6)	Journals	856	862	866	860	850	856	868	887	907	956	1,006	1,062	1,112	1,169	1,213	1,221
	Articles	77,896	76,718	75,342	73,100	69,700	64,480	68,572	70,960	73,467	79,348	86,516	94,518	102,304	109,886	113,928	114,774
	Article/Journals	91	89	87	85	82	80	79	80	81	83	86	89	92	94	94	94
(7)	Journals	76	78	83	91	101	114	124	134	140	142	146	156	166	179	193	204
	Articles	4,280	4,469	4,819	5,188	5,652	6,275	6,696	7,434	7,904	8,158	8,240	8,564	8,943	10,017	11,378	12,401
	Article/Journals	56	57	58	57	56	55	54	56	56	57	56	55	56	56	59	61
(8)	Journals	903	893	886	893	901	900	927	966	1,030	1,094	1,156	1,213	1,269	1,330	1,351	1,340
	Articles	43,344	41,971	41,642	41,078	37,842	35,100	36,153	40,572	47,380	50,324	48,552	49,733	50,760	53,200	54,040	53,600
	Article/Journals	48	47	47	46	42	39	39	42	46	46	42	41	40	40	40	40
(9)	Journals	252	254	254	254	254	256	272	272	280	288	296	298	298	300	302	304
	Articles	12,652	13,217	13,661	14,299	15,033	15,818	19,056	24,921	32,178	39,192	41,634	39,611	37,253	38,944	41,763	42,091
	Article/Journals	50	52	54	56	59	62	70	92	115	136	141	133	125	130	138	138
(10)	Journals	2,815	2,830	2,854	2,897	2,947	3,010	3,109	3,221	3,349	3,505	3,656	3,804	3,935	4,078	4,164	4,175
	Articles	208,320	208,816	210,998	213,635	213,486	217,424	229,907	251,813	275,952	295,453	305,420	318,427	329,733	342,747	350,933	353,718
	Article/Journals	74	74	74	74	72	72	74	78	82	84	84	84	84	84	84	85

	1976	1977	1978	1979	1980	1981	1982	1983	1984	1985	Percent Change 1960–1965	1965–1970	1970–1975	1975–1980	1980–1985
Journals	274	278	280	282	284	286	288	290	292	294	45	30	9	5	4
(1) Articles	56,419	57,485	57,614	57,871	58,223	58,647	59,124	59,641	60,188	60,757	48	21	8	4	4
Article/Journals	206	207	206	205	205	205	205	206	206	207	3	-8	-1	-1	1
Journals	102	106	110	114	118	122	126	130	134	138	21	30	18	20	17
(2) Articles	8,364	8,586	8,846	9,086	9,304	9,499	9,669	9,813	9,929	10,015	53	75	10	14	8
Article/Journals	82	81	80	80	79	78	77	75	74	73	27	35	-7	5	8
Journals	65	68	71	74	77	80	83	86	89	92	50	39	24	24	19
(3) Articles	8,580	8,976	9,202	9,464	9,747	10,043	10,347	10,941	11,234	11,224	43	28	19	15	15
Article/Journals	132	132	130	128	127	126	125	124	123	122	7	3	3	4	4
Journals	231	234	237	240	243	247	250	253	256	259	17	15	9	7	7
(4) Articles	15,939	17,016	18,293	19,195	20,447	21,677	22,583	23,834	25,064	26,296	35	58	-5	39	32
Article/Journals	69	73	77	80	84	88	90	94	98	102	15	37	-12	29	21
Journals	460	471	473	478	484	491	497	504	509	516	14	12	-3	8	7
(5) Articles	45,080	46,629	47,395	48,419	49,665	51,066	52,575	54,159	55,816	56,863	7	30	12	13	14
Article/Journals	98	99	100	101	103	104	106	107	110	110	6	16	15	5	7
Journals	1,268	1,318	1,359	1,400	1,441	1,482	1,523	1,563	1,606	1,648	0	18	21	18	14
(6) Articles	121,728	127,846	133,410	138,914	144,382	149,829	155,263	160,690	166,393	171,978	-12	26	33	26	19
Article/Joucnals	96	97	98	99	100	101	102	103	104	104	12	8	9	6	4
Journals	216	228	239	251	263	275	287	300	313	326	50	28	40	29	24
(7) Articles	13,307	13,425	14,162	14,926	15,707	16,475	17,233	18,011	18,777	19,534	47	28	40	29	24
Article/Journals	62	59	59	59	60	60	60	60	60	60	-2	2	9	-2	0
Journals	1,383	1,428	1,455	1,483	1,512	1,542	1,572	1,606	1,641	1,678	0	28	16	12	11
(8) Articles	55,320	57,120	57,695	58,183	58,603	58,979	59,320	59,626	59,913	60,179	81	38	10	9	3
Article/Journals	40	40	40	39	39	38	38	37	37	36	19	8	5	-2	-8
Journals	310	316	321	326	331	336	340	344	349	353	2	16	3	9	7
(9) Articles	43,710	45,188	46,972	48,780	50,607	52,449	54,303	56,165	58,035	59,911	25	163	1	20	18
Article/Journals	141	143	146	150	153	156	160	163	166	170	24	127	-2	11	11
Journals	4,309	4,447	4,545	4,648	4,753	4,861	4,966	5,076	5,189	5,304	7	21	18	14	12
(10) Articles	363,447	382,271	393,589	404,838	416,685	428,664	440,417	452,585	465,056	476,767	4	40	16	21	14
Article/Journals	86	86	87	87	88	88	89	89	90	90	-3	17	1	4	2

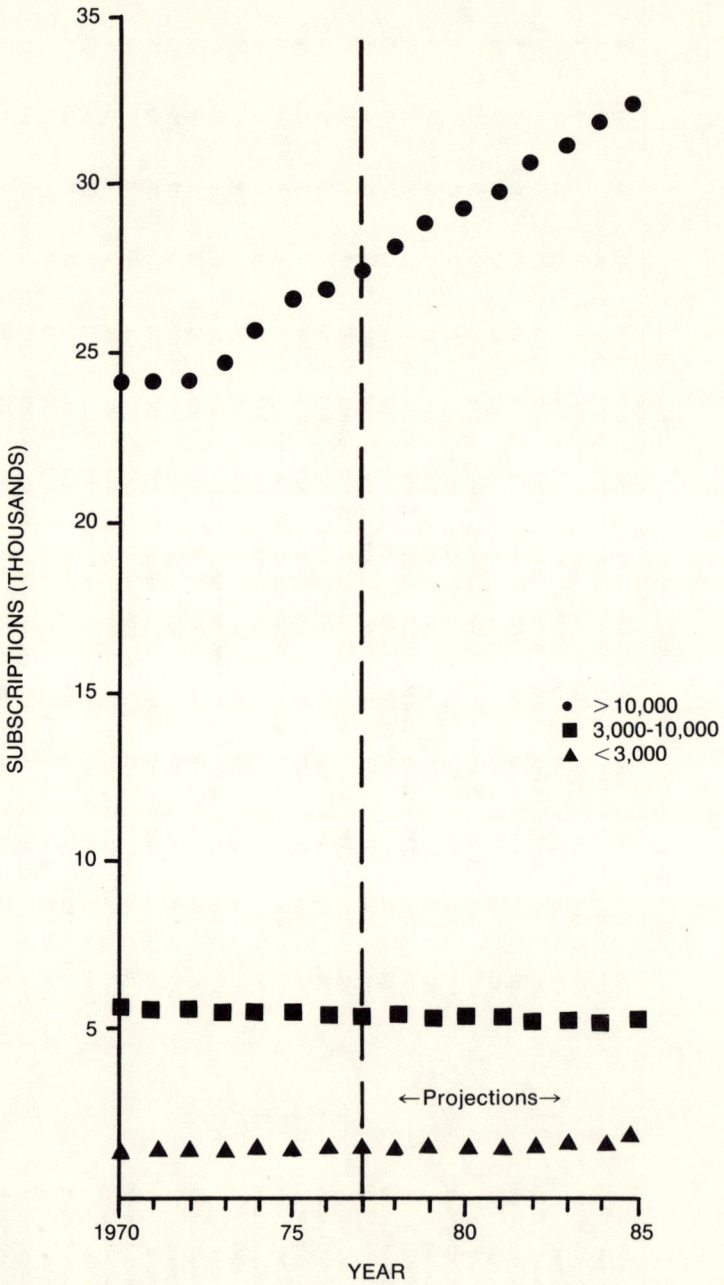

Figure 4.2 Growth of the number of subscriptions by size of journals, 1970–1985.

GENERAL CHARACTERISTICS OF U.S. SCIENTIFIC AND TECHNICAL PERIODICAL PUBLISHING

Types of Publishers

One striking characteristic of the scientific and technical periodical publishing industry is the wide variety of types of publishers involved. Publishers are the system participants responsible for editing, reproducing, and initially distributing scientific and technical periodicals. Some professional societies and other institutions publish their periodicals by proxy, that is, they engage publishers to perform editorial and/or other functions. And some publishers do not set type or print copies, but rather subcontract to other organizations for such services.

Scientific and technical periodical publishers may be classified as follows:

1. Societies and associations include both large and small professional and trade associations (such as Association for Computing Machinery, American Chemical Society, and American Genetic Association).
2. Commercial publishers include general scientific and professional publishers and publishers that concentrate on special subject areas (such as Academic Press, John Wiley & Sons, and Williams & Wilkins).
3. Educational institutions are university presses, or universities, colleges, technical schools, and other institutions of learning (such as Harvard University Press and University of Chicago Press).
4. Other or undetermined publishers including government agencies, corporations that are not primarily publishers, research or nonprofit organizations, or other undetermined organizations.

The estimated numbers of scientific and technical periodicals handled by these types of publishers are given in Table 4.4. Societies and associations account for the largest proportion of periodicals, whether they be journals or other periodicals. Many societies and associations publish both a scholarly journal and a popular newsletter or bulletin. A high proportion of periodicals that are other or undetermined are government periodicals or trade journals published by companies such as IBM or Control Data Corporation.

Table 4.4   Number of U.S. scientific and technical periodicals, by type of publisher, 1977.

Type of Publisher	Total Periodicals	Pro- portion (%)	Journals	Pro- portion (%)	Other Periodicals	Pro- portion (%)
Societies and associations	3,387	38	1,734	39	1,653	37
Commercial publishers	2,629	29	1,557	35	1,072	24
Educational institutions	1,247	14	800	18	447	10
Other or undetermined	1,652	19	356	8	1,296	29
All types	8,915	100	4,447	100	4,468	100

Source: King Research, Inc. Journal Tracking Survey

Language of Publication and Address of Publishers

One indication of the amount of transnational flow of scientific and technical journals is the language of journals and multinational addresses of publishers. Data on language of publication and publisher address were recorded for a sample of scientific and technical journals coded as "U.S." by *Ulrich's*.[10] The proportions falling into each of these categories were then projected to the population of 4,447 U.S. S&T journals in 1977 for each field of science. (Table 4.5).

The first column of Table 4.5, "English Text," shows, as expected, that the majority of S&T journals published in the United States have an original English text (4,277 out of 4,445 or 96%). Much smaller proportions of U.S. journals publish non-English or polyglot text (38 or 0.8%) and a small proportion are translations into English from foreign languages (130, or 3%). Life Sciences, Psychology, Social Sciences, and Other Sciences showed no translation journals published in the United States, while Physical Sciences has the highest proportion (18%).

The "Publisher Address" columns in Table 4.5 give the number of U.S. journals published by companies with offices both within and outside the U.S., based upon observation of these categories in *Ulrich's*[11] for the journal tracking survey sample. Over all fields of science, 4,129 U.S. scientific and technical journals list only a U.S. publisher address (93%). Only 316 (7%) of all U.S. S&T journals list publishers with both U.S. and foreign addresses. These proportions vary, however, for individual fields of science. No Social Sciences journal lists both U.S. and foreign publisher addresses, while a very high

Table 4.5 Language and address of publishers of U.S. scientific and technical journals, by field of science, 1977.

Field of Science	Language of Publication			Publisher Address		Total Journals
	English Text	Non-English or Polyglot[a] Language Text	Translation into English	U.S.	U.S. and Non-U.S.	
Physical Sciences	214	23	51	176	112	288
Mathematics	88	7	6	79	22	101
Computer Sciences	54		10	55	9	64
Environmental Sciences	202	4	17	212	11	223
Engineering	430		46	444	32	476
Life Sciences	1,302			1,177	125	1,302
Psychology	207	4		207	4	211
Social Sciences	1,474			1,474		1,474
Other Sciences	306			305	1	306
All Fields	4,277	38	130	4,129	316	4,445[b]

Source: King Research, Inc. Journal Tracking Survey
[a]Polyglot journals include articles in different languages; for example, some in English and some in French.
[b]Do not sum to 4,447 because of rounding.

proportion of Physical Sciences journals (112, or 39%) do so. This may demonstrate that the Physical Sciences are more "international" in nature than the other fields of science, reflected by publishers that maintain editorial, marketing, and/or distribution facilities both within and outside the United States.

Age of U.S. Scientific and Technical Periodicals

The average age of periodicals gives one indication of the dynamism of publishing. Average age was estimated by field of science and type of publisher. The first year of publication for a sample of periodicals was recorded from *Ulrich's,*[12] and an average age in years was calculated as of 1975 (see Tables 4.6 and 4.7). The data do not take into account the fact that two distinct periodicals may derive from a common progenitor (for example, *Industrial and Engineering Chemistry*

Table 4.6 Average age of U.S. scientific and technical periodicals, by field of science, 1975 (in years).

	Journals	Other Periodicals	All Periodicals
Physical Sciences	18.5	15.5	17.5
Mathematics	24.8	10.4	23.0
Computer Sciences	14.1	8.6	11.3
Environmental Sciences	28.0	15.7	21.2
Engineering	21.6	33.4	29.1
Life Sciences	31.2	18.9	26.2
Psychology	18.6	4.4	16.3
Social Sciences	23.4	13.6	17.8
Other Sciences	29.4	39.0	34.1
All Fields	25.4	20.0	22.7

Table 4.7 Average age of U.S. scientific and technical periodicals, by type of publisher, 1975 (in years).

Type of Periodical	Commercial Publisher	Society or Association	Educational Institution	Other Publishers	All Types
Journals	16.0	29.1	28.6	N.A.	25.4
Other Periodicals	24.5	18.3	27.7	29.4	20.0
All Periodicals	19.8	24.0	28.4	29.4	22.7

and *Chemical and Engineering News*); one of the second generation periodicals is often considered to be a continuation and the other a new periodical. Over all fields of science, U.S. scientific and technical periodicals had an average age of 22.7 years in 1975. Journals were somewhat older than other periodicals, 25.4 and 20.0 years, respectively. Except in Engineering and Other Sciences, other periodicals are generally not as old as journals, perhaps indicating that news-oriented bulletins, newsletters, and trade journals either have shorter life spans, or are a newer entry to the scientific and technical periodical publishing scene. Interpretation of age data is limited by the fact that systematic data are lacking on defunct periodicals.

Average age of periodicals differ significantly among the fields of science. Other Sciences, which includes such multidisciplinary publications as *Scientific American* (started 1845) and *Science* (started 1880) average 29.4 years for journals, 39.0 years for other periodicals, and 34.1 years for all scientific and technical periodicals. Computer Science journals average 14.1 years, evidence of the relative newness

Table 4.8 Local and regional U.S. scientific and technical journals, by field of science, 1977.

Field of Science	Local or Regional		National		Total Number
	Number	Proportion (%)	Number	Proportion (%)	
Physical Sciences			288	100	288
Mathematics			104	100	104
Computer Sciences			66	100	66
Environmental Sciences	34	14	210	86	244
Engineering			476	100	476
Life Sciences	286	22	1,016	78	1,302
Psychology	4	2	214	98	218
Social Sciences	270	19	1,153	81	1,423
Other Sciences	16	5	310	95	326
All Fields	610	14	3,837	86	4,447

Source: King Research, Inc. Journal Tracking Survey.

of this field when compared with Life Sciences (31.2 years), Environmental Sciences (28.0 years) and Mathematics (24.8 years).

Geographic Coverage of U.S. Scientific and Technical Journals

Journals in the sample were classified as local or regional according to whether they appeared to be sponsored by a local, state, or multistate organization, and/or if they were distributed only in a local, state, or multistate region. The rest were classified as national publications and include journals with substantial foreign distribution. A few journals known from external evidence to have national distribution (such as *New England Journal of Medicine*) were classified as national journals even though their names imply a regional level journal (see Table 4.8).

Over all fields of science, approximately 86 percent (3,837) of all journals are national, with 14 percent (610) regional. Our sample included no regional journals in Physical Sciences (the *Astronomical Society of the Pacific Journal* was classified as national), Mathematics, Computer Sciences, or Engineering. Life Sciences, Social Sciences, and Environmental Sciences have the highest proportion of regional journals, with 22 percent, 19 percent, and 14 percent, respectively.

Environmental Sciences has a number of journals that focus on regional environmental concerns. In Life Sciences, a number of state medical societies publish their own journals. A number of the Social Sciences regional journals are university press publications.

ECONOMIC CHARACTERISTICS OF U.S. SCIENTIFIC AND TECHNICAL JOURNALS

Costs of Publishing U.S. Scientific and Technical Journals

The dramatically increasing costs of publishing scientific and technical journals have been a major concern of journal publishers over the past two decades. The trend in the total cost of journal publishing is displayed in Figure 4.3 in current and constant dollars.* It is clear that, even in constant dollars, the total cost of publishing has been increasing steadily since 1960. Costs are expected to continue to rise through 1985. The elements of cost are described in some detail to suggest the many factors involved in the noted increases.

In addition to the resources that go into publishing (labor, equipment, supplies, postage, and support facilities), a significant influence on costs is the fact that journals currently publish more articles now than in the past. In 1960 journals averaged 68 articles per journal; this number increased to 86 in 1977 (see Figure 4.4). In fact, increases in cost per article are less than the increases in cost per journal.

The number of subscriptions distributed by publishers is very important because of the structure of publishing costs. Journal publishing is characterized by large costs that are incurred in preparing the master copy of a publication. These costs, which are incurred for such activities as editing, typesetting, and redacting, are referred to as prerun costs. The costs of reproducing and distributing additional copies are called runoff costs. The runoff costs are incurred for such activities as printing, collating, binding, and mailing.

As the number of subscriptions increases, runoff costs and thus total expenditures for journal publication increase. The average cost, however, decreases. In 1977, for example, the per subscription publisher expenditures associated with small journals were $60, while the corresponding figures for medium and large journals were $29 and $11,

---

*1972 constant dollars are calculated using the GNP implicit price deflator.

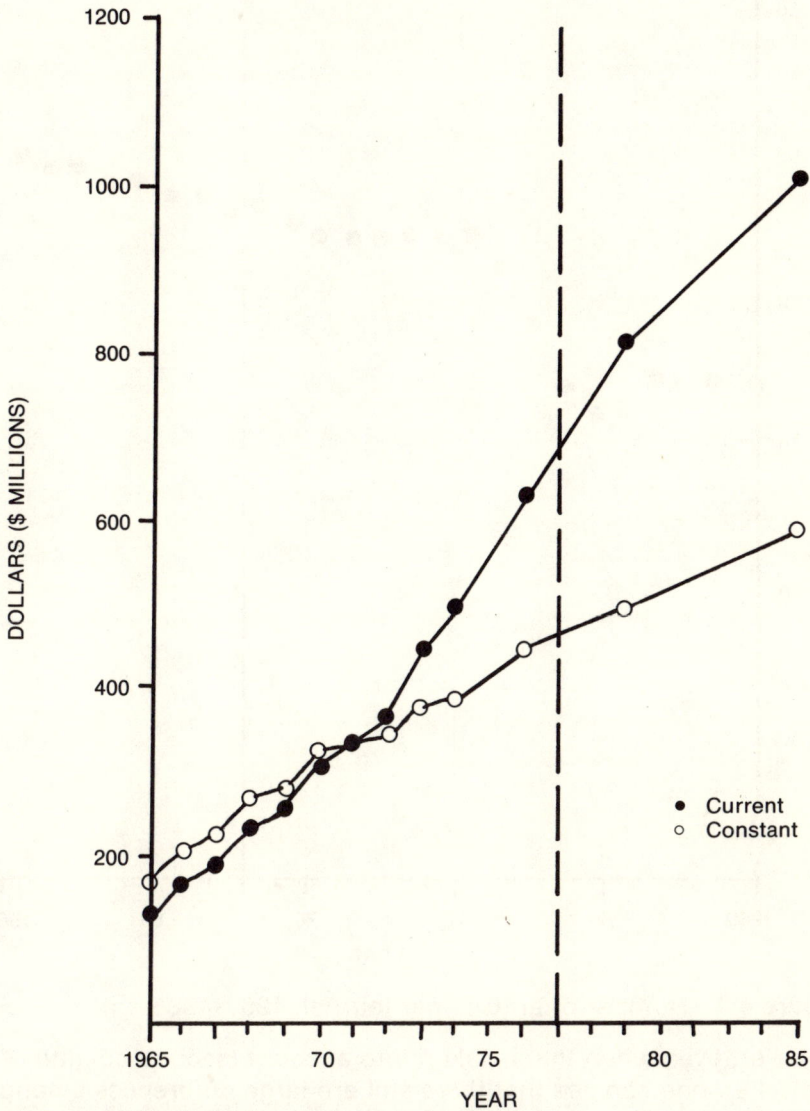

Figure 4.3   Total cost of journal publishing in current and constant dollars, 1965–1985.

respectively. The average cost per journal also varies somewhat among fields of science. (see Table 4.9). The differences observed in average cost among fields of science are partly attributable to the fact that the average number of subscriptions varies somewhat among the fields.

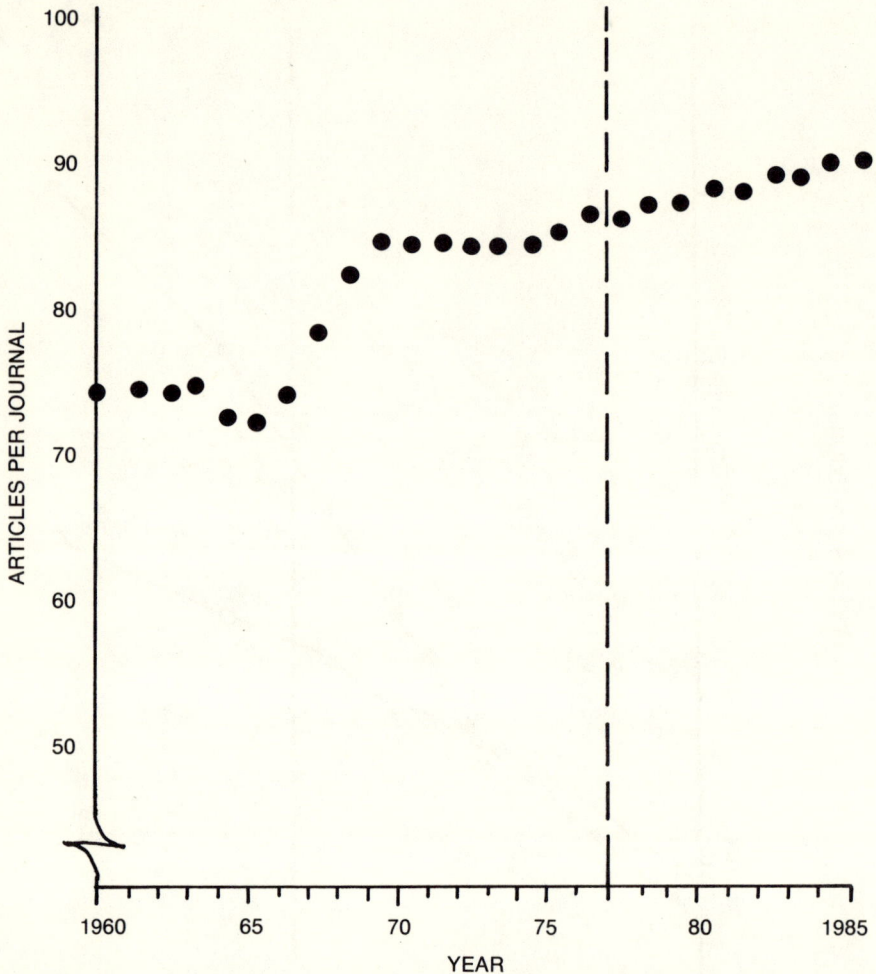

Figure 4.4   Number of articles per journal, 1960–1985.

However, even when that is taken into account (second column of Table 4.9), one can see that there still are large differences among fields. These differences are partially accounted for by differences in the numbers of pages published, as shown in the last column of Table 4.9. The remaining cost differences can be explained by page makeup and typesetting costs. Articles in some fields, such as Physical Sciences and Mathematics, have a large number of special graphics (chemical compound symbols, mathematical formulae, and line graphs). The cost of preparing these graphics is very high. In addition, the number of words per page will also affect typesetting costs.

Table 4.9 Cost per journal and subscription, by field of science, 1977.

Field of Science	Cost per Journal	Cost per Subscription	Cost per Page Distributed
Physical Sciences	$206,700	$57.40	$0.037
Mathematics	125,400	28.80	0.029
Computer Sciences	201,100	11.70	0.015
Environmental Sciences	148,700	12.80	0.017
Engineering	136,800	16.10	0.020
Life Sciences	123,200	18.60	0.022
Psychology	70,000	20.10	0.036
Social Sciences	65,600	28.20	0.045
Other Sciences	297,100	16.30	0.011
All Fields	120,400	19.00	0.023

Prices of Scientific and Technical Journals

The costs of journal publishing obviously have some impact on the prices that are charged for journals. Thus the trends in cost should be reflected in price data, and the differences in cost observed by size of journal and among fields of science should also be reflected in the price data for the corresponding journal categories.

Journal publishers often charge varied prices to different subscribers. The principal distinctions among subscribers are made by journals published by societies and associations in which scientists and engineers are members. Members are usually charged less, and the price is sometimes made a part of membership dues. Another distinction is whether the subscriber is an individual (without attention to membership), an institution (library), or foreign. Estimated prices for these five categories of subscribers are shown in Table 4.10 for 1975 and 1977. The lowest average price is paid by individual members ($16 in 1977), followed by individual nonmembers ($26 in 1977). For those journals that make no distinction among individuals, the average price charged is about the same as the price charged to institutions ($38 in 1977). Foreign subscribers are charged an average of $42. The higher price to foreign subscribers partially reflects the extra cost of mailing overseas, and distinctions are sometimes made among foreign countries.

When all individual prices are combined, the trends of the three remaining price categories appear as given in Figure 4.5. Differences in trends by type of subscriber due to changes in cost are not expected, since such costs do not differ much. On the other hand, publishers

Table 4.10 Journal subscription price for five categories of subscribers, 1975 and 1977.

Subscriber Category	1975	1977
Member	$13.54	$16.09
Nonmember	23.80	26.04
Individual (no membership distinction)	30.87	37.94
Institution	30.47	37.72
Foreign	33.37	42.05

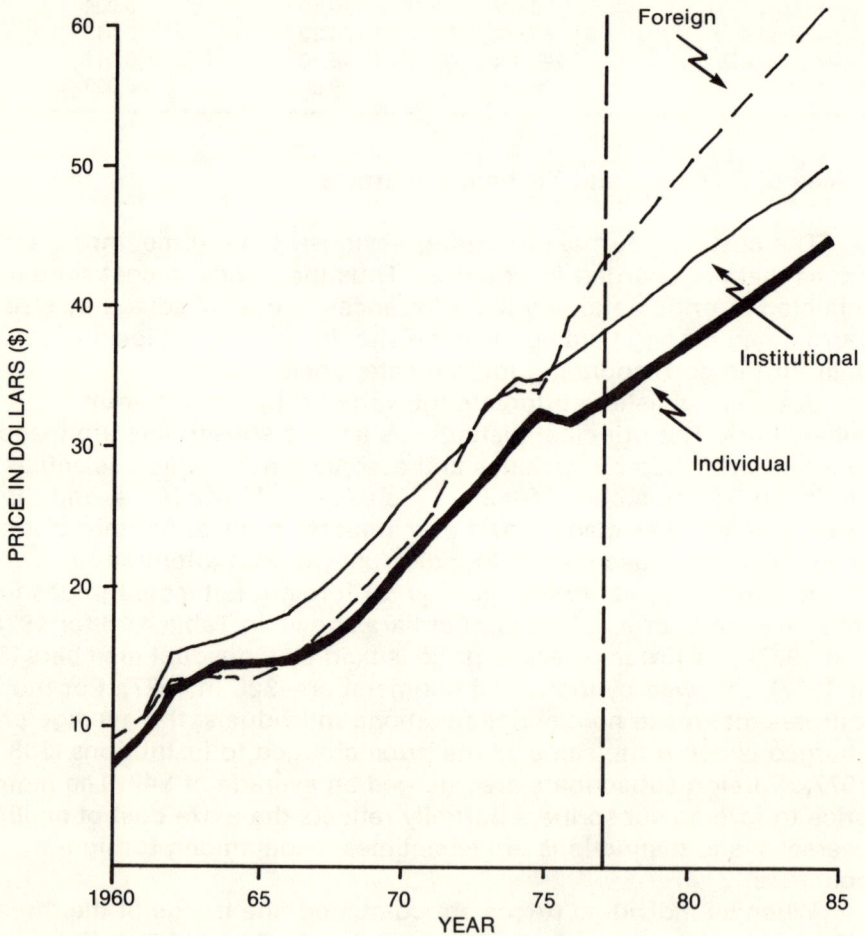

Figure 4.5 Journal subscription prices in current dollars, 1960–1985.

Table 4.11   Price per journal, by field of science, 1977.

Field of Science	Average Price per journal
Physical Sciences	$96.83
Mathematics	56.44
Computer Sciences	46.30
Environmental Sciences	30.73
Engineering	53.35
Life Sciences	37.69
Psychology	24.85
Social Sciences	19.42
Other Sciences	20.42
All Fields	35.91

recognize that the price elasticities of the different groups are likely to be different and do charge accordingly. However, the increase in price is not much different among individual, institution, and foreign subscriptions. From 1960 to 1977, individual subscription prices rose 317 percent, institutional prices 306 percent, and foreign prices 348 percent.

The 1977 average prices are given by field of science in Table 4.11. As indicated, there is a considerable range, with Social and Other Sciences journal prices averaging about $20 and Physical Sciences journal prices nearly $100. The estimated average price per journal also varies according to the number of subscriptions. The average price for small subscription journals is $41.62, for medium journals $31.83, and for large journals $21.47. This partially reflects the similar differences in cost per journal subscription.

Number of Scientific and Technical Journal Subscriptions

Another important variable in the cost of scientific and technical journal publication is the number of subscriptions. The estimated average number of subscriptions per U.S. journal was 6,327 in 1977. Over one-half of these subscriptions are estimated to be U.S. individuals (3,542). U.S. institutional subscriptions are estimated to be 1,789, and foreign subscriptions 996. The trends in these types of subscriptions vary substantially as shown in Figure 4.6.* Since 1965 the number of

*Subscription data were obtained from a variety of sources. The more recent data are more reliable.

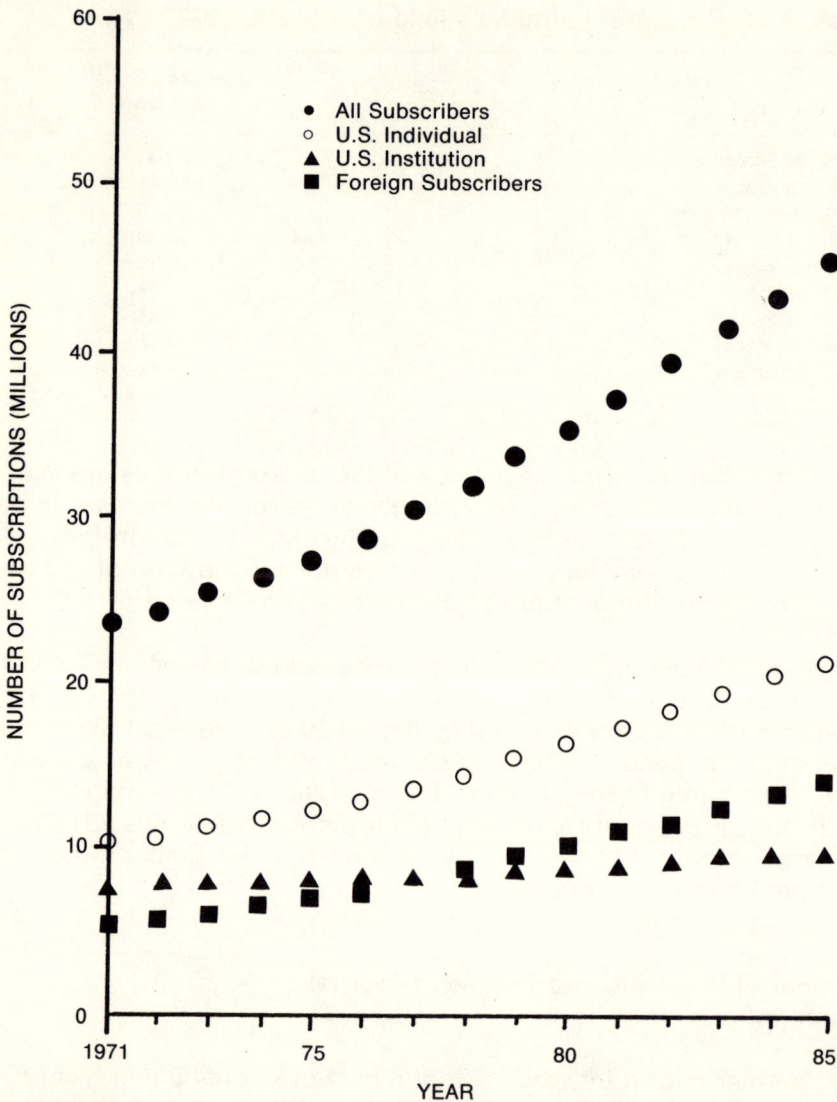

Figure 4.6 Total number of subscriptions by type of subscribers, 1971–1985.

individual subscriptions per journal has increased about 13 percent and foreign subscriptions have increased about 11 percent. Institutional subscriptions, however, show a decrease of about two percent overall. The figure includes projections of subscriptions to 1985.

The distribution of number of subscriptions per journal is very highly skewed; that is, there are many small journals and a few very large journals (see Table 4.12). To illustrate the differences among the different sizes of journals, we have grouped them into three categories: small (less than 3,000 subscribers), medium (3,000–10,000 subscribers), and large (over 10,000 subscribers). The average number of subscriptions for small journals is estimated to be 1,391; for medium journals, 5,454, and for large journals, 27,450. Since many of the large journals have a very large number of subscriptions, they dominate the total number of subscriptions even though there are fewer of them (see Table 4.13) Even though only 13 percent of the journals are classified as large they account for 58 percent of all subscriptions; but the one-half of the journals that are considered small contribute only 11 percent of the subscriptions.

Although there has been steady growth in the total number of subscriptions, as shown in Figure 4.7, the growth appears to be largely accounted for by increases in number of scientists and engineers. This is made clear by data given in Table 4.14. Between 1965 and 1977, the average number of subscriptions increased from 9.5 to 10.3 and the number of individual subscriptions per scientist or engineer has increased only slightly from 5.0 to 5.7. Comparable data for 1977 by field of science, shown in Table 4.15, indicate a range of subscriptions per scientist or engineer of 1.1 (Physical Sciences) to 24.3 (Life Sciences). The figure for the Life Sciences is artificially high because our estimated number of scientists in that field excludes medical practitioners, who are of course subscribers as well. The high figure in the Environmental Sciences probably reflects the multidisciplinary nature of the field, with subscribers drawn from several fields of science.

Income for Scientific and Technical Journal Publishing

Most journals get the great bulk of their income from subscriptions, and, in certain cases, from page charges. Occasionlly there is a significant subsidy from a society or institution sponsoring a journal. A few large journals get a sizable part of their income from advertising. Other minor sources of income include royalties, endowments, and sales of reprints, back issues, and microforms. The proportion of each type of income will vary significantly from journal to journal, depending on factors such as the type of publisher and the size of the journal.

Subscription income is that received from the various groups of subscribers including individuals, institutions, and foreign. Average 1977

Table 4.12 Number of journals by subscription size categories, 1977.

	0–1,000	1,001–2,000	2,001–3,000	3,001–4,000	4,001–5,000	5,001–10,000	10,001–15,000	15,001–20,000	20,001–50,000	50,001–100,000	Over 100,000	Total
						Number of Subscriptions						
Number of Journals	978	786	507	399	499	682	149	60	194	104	89	4447
Proportion of Journals	22%	18%	11%	09%	11%	15%	3%	1%	4%	3%	2%	100%

Source: King Research, Inc. Publisher Survey; *Ulrich's Directory of International Periodicals.*

Table 4.13   Total number of subscriptions by size of journals, 1977.

Size of Journal	Number of Journals	Average Number of Subscriptions	Total Number of Subscriptions (millions)
Small	2,271	1,391	3.2
Medium	1,580	5,454	8.6
Large	596	27,450	16.3
Total	4,447	6,327	28.1

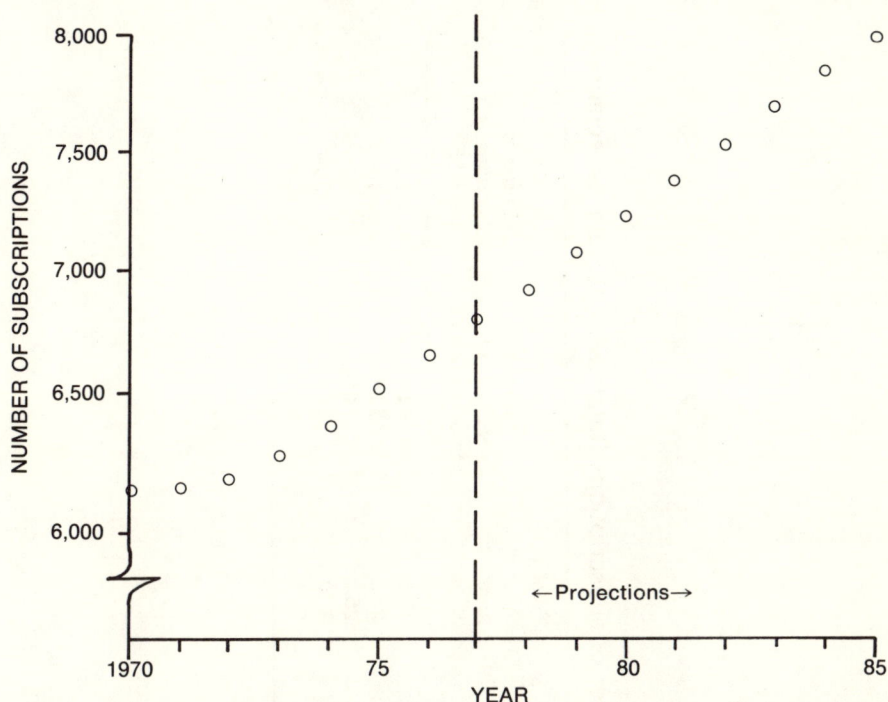

Figure 4.7   Growth of number of journal subscriptions, 1970–1985.

subscription income for journals derived from these estimates is given by field of science in Table 4.16. An alternative way of deriving 1977 subscription income is to develop journal publication costs, and, based on identified relationships between publication costs and subscription income, calculate subscription income. This approach, which we have adopted in our publication cost model, yields the second column of results in Table 4.16. These two sets of results can be viewed as bounds on the actual subscription income. As indicated previously, income will also vary substantially by size of journal (see Table 4.17).

Table 4.14 Average number of subscriptions (total, average, and per scientist and engineer), 1977.

Year	Number of Journals	Total Number of Subscriptions (millions)	Average Number of Subscriptions	Number of Subscriptions per S&E	Number of Individual Subscriptions (millions)	Individual Subscriptions per S&E
1965	3,010	17.6	5,862	9.5	9.4	5.0
1970	3,656	23.2	6,340	9.7	12.3	5.2
1975	4,175	25.3	6,056	9.6	13.5	5.1
1976	4,309	27.0	6,259	10.0	14.9	5.5
1977	4,447	28.1	6,327	10.2	15.7	5.7

Table 4.15 Number of subscriptions per journal, total number of subscriptions per scientist or engineer, and individual subscriptions per scientist or engineer, by field of science, 1977.

Field of Science	Number of Subscriptions per Journal	Total Number Subscription (000)	Subscriptions per S&E	Individual Subscriptions (000)	Individual Subscriptions per S&E
Physical Sciences	3,727	1,036	3.61	327	1.1
Mathematics	4,276	453	4.04	258	2.3
Computer Sciences	16,719	1,137	5.92	452	2.3
Environmental Sciences	12,077	2,826	31.40	1,730	19.2
Engineering	8,573	4,038	2.93	1,637	1.2
Life Sciences	6,539	8,619	26.68	5,628	24.3
Psychology	3,330	759	5.58	263	1.9
Social Sciences	2,316	3,307	13.55	1,717	7.0
Other Sciences	18,860	5,960	N.A.	3,538	N.A.
All Fields	6,327	28,135	10.19	15,750	5.7

Table 4.16 Average journal subscription income by field
of science, 1977.

	Subscription Income Per Journal	
Field of Science	Based on Subscription Prices	Based on Estimated Costs
Physical Sciences	$232,300	$191,100
Mathematics	164,000	120,000
Computer Sciences	531,800	184,300
Environmental Sciences	237,300	121,200
Engineering	301,700	106,600
Life Sciences	166,400	99,200
Psychology	57,700	69,900
Social Sciences	30,100	72,000
Other Sciences	248,800	180,200
All Fields	151,400	109,600

Table 4.17 Average subscription income, by size of journal, 1977.

Size of Journal	Based on Subscription Prices	Based on Estimated Costs
Small	$ 69,200	$ 70,100
Medium	164,800	124,600
Large	472,400	241,100
Total	151,400	109.600

Subscription income as a proportion of total publication income can
vary significantly among journals. In 1968, data on a number of society
journals were analyzed by Conyers Herring for the Report of the Task
Group on the Economics of Primary Publication.[13] This study indicated a
range of from less than 20 to nearly 100 percent of total income derived
from subscriptions. Herring speculated that this proportion might vary
with circulation, with income coming predominantly from page charges
for journals with low circulation, from advertising for journals with very
high circulation, and from subscriptions for the remainder of the
journals.

The sources of income are given for two types of publisher in
Table 4.18. These data show each group to have average subscription
income of 86 percent, indicating that most journals fall in Herring's
mid-range, where subscriptions are the major source of income. The
breakdown by type of publisher is derived from a sample of 167 journals
in both the natural and social sciences.

Table 4.18   Proportion of publishing income, by type of publisher, 1977.

Income Source	Type of Publisher		Total
	Professional Society	Commercial	
Subscription[a]	86.3%	86.3%	81.1%
Advertising	3.3	4.2	4.5
Back Issues, reprints, and microform	4.5	9.5	7.8
Page charges	5.9		6.6
Total	100.0	100.0	100.0

Source: K. W. Leeson and F. Machlup, "A Catalog of Statistical Data on the Economic Characteristics of Primary Research Journals in the Sciences and Technology," Preliminary Findings, March 18, 1977.
[a]Includes miscellaneous income, such as society subsidies.

   The proportion of subscription income in relation to other forms of journal income for the nine fields of science is presented in Table 4.19. An average of about 81 percent of income comes from subscriptions, ranging from 71 percent for the Life Sciences to 97 percent for Social Sciences.

   Two sets of data for advertising income are shown in Tables 4.18 and 4.19. Many journals carry advertising; in two early studies of scientific and technical journals,[14,15] about half were found to do so. Only rarely, however, does the net income from advertising support a substantial part of the cost of publishing. The exceptions are some journals with very large circulations, where yield from advertising can be very high, and news-type journals, where advertising may form an integral part of the current awareness function of the journal. The proportion of income received from advertising may range up to 75 percent or more for individual publishers. However, on the average it is somewhat less than 5 percent. The proportion of journals carrying advertising and the amount of advertising income varies substantially by field, with Life and Environmental Sciences journals carrying more advertising than average. Commercial journals receive about twice as much of their income from advertising as do society or association publishers.

   A small part of journal income, usually less than 10 percent, is derived from the sale of reprints, back issues, and microform copies of the journal. Back-issue sales are generally made to libraries, with more sales by commercial publishers, probably because of the frequent

Table 4.19 Proportion of publishing income received from various sources, by field of science, 1977.

Income Source	All Fields	Physical Sciences	Mathematics	Computer Sciences	Environmental Sciences	Engineering	Life Sciences	Psychology	Social Sciences	Other Sciences
Subscriptions[a]	81%	82%	85%	81%	72%	69%	71%	89%	97%	54%
Advertising	5	2	6	10	18	15	16	7	1	43
Back issues, reprints and microform	8	10	8	8	8	3	10	2	2	2
Page charges	7	6	1	1	2	13	3	2	2	1
All sources	100	100	100	100	100	100	100	100	100	100

[a]Includes miscellaneous income, such as society subsidies.

availability of back issues of society journals from members. Reprint sales account for more income than back issues and microforms combined, with some journals having well-publicized programs for reprint sales to the general public. Some reprints are tied in with payment of page charges. Reprints are also sold, sometimes in large quantities, to article authors.

Page charges, another source of publisher income, can account for an appreciable proportion of the total. When charged (most prominently in the Physical Sciences and Engineering and only by not-for-profit journals), the intent is generally to cover all costs associated with the production of a journal article up to the actual printing (the prerun costs), which may range from less than 25 to more than 65 percent of total costs, with the proportion for the average journal being about half. Although page charges are often set somewhat below prerun costs and are not collected from all authors, they may still account for more than half of the total income.[16] Overall these charges make up 20 percent of the income of those journals utilizing them and about 7 percent of the combined income of all journals. By field of science, the percentage of total income attributable to page charges ranges from 13 percent for Engineering to only a trace for the Social Sciences.

A COST MODEL FOR JOURNAL PUBLISHING

Description of Journal Publishing Activities

Costs incurred in publishing can be categorized in a number of ways, but they are most commonly grouped into prerun, runoff, and miscellaneous or other costs. Specific activities related to these three categories are given in Table 4.20. Prerun operations are those necessary before production of the first copy of the printed work. Related costs are clearly dependent on the number, size, and complexity of the manuscripts submitted. Runoff costs depend largely on the number of subscribers and, to a lesser extent, on the number of pages. Miscellaneous activities are generally viewed as incidental to the publication process. The model accounts for costs incurred in publishing, not in writing or rewriting the manuscripts or in gathering the underlying information.

The activities listed in Table 4.20 can be performed in a number of locations and by a number of individuals, not all of are necessarily paid by the publisher. Reviewing, and sometimes subject editing, are usually performed by individual scientists working at their regular location;

Table 4.20 Activities performed in journal publication.

Prerun	Runoff	Miscellaneous
Subject editing	Subscription maintenance	Promotion
Review[a]	Printing	Advertising
Copy editing	Binding	Conversion to microform
Graphics preparation	Wrapping and mailing	Reprint and back issue
Typesetting		sales
Engraving		
Proofreading		

[a]Peer review or refereeing.

editing and production tasks are often conducted in separate locations; and typesetting and printing are frequently contracted out.

In this chapter, each activity in journal publishing is defined in terms of input, process, and output. Three sources of data are incorporated in the model: journal characteristics, process variables, and cost elements. The input, process, and output associated with each activity are briefly described.

Editing: Staff

Input: Manuscript submitted to journal.
Process: Review by editorial staff, submission to reviewers, decision to accept or reject, editing.
Output: Manuscript accepted and ready for composition.

Once a manuscript has been submitted to a journal, it goes through an extensive review process, which usually involves the managing editor, a subject editor, and one or more reviewers. The manuscript is routed through each of these stages in turn for evaluation and review. At any stage, the decision may be made to reject the manuscript, in which case the author is notified and the process terminated. When a manuscript has been accepted conditionally, it will be returned to the author for modifications and then reenter the review cycle. After final acceptance, final editing is performed on the manuscript.

This activity category covers that portion of the editorial processes performed by the publisher's staff. The major staff participants in the editing process, which is highly labor intensive, are the managing editor, subject editor or editors, and secretaries and clerks. Other costs incurred are those associated with the various mailings to reviewers, authors, and so on, and also general overhead expenses.

In the publication cost model, the level of each activity is determined by tracking manuscripts through the processing stages. Important determinants of the level of activity are the number of manuscripts submitted and the number accepted, accepted with modification, or rejected at each stage. Editorial costs are computed for an individual, average journal, so that the costs of processing rejected manuscripts are included and eventually are allocated to those manuscripts that are published. Also included is the cost of preparing nonarticle material, such as editorials and book reviews. Since we present only U.S. uses of the journal literature, editing costs are prorated to U.S. and foreign subscribers and only the U.S. portion is included in the cost model. This approach ignores the costs attributable to U.S. subscriptions to foreign journals.

To compute total costs for editing, per-journal editing costs are multiplied by the number of journals published. This represents the total cost of editing.

Editing: Donated

Input: Manuscript submitted to journal and transmitted to subject editor.
Process: Review by subject editor, selection of reviewers, decision to accept or reject.
Output: Accepted manuscript.

This category is part of the general editing activity. For some society journals, the subject editors are scientists who are not part of the regular publishing staff and are paid at most a small stipend. This category accounts for the difference between the economic cost of a paid subject editor (and secretary) and the stipend actually paid to a volunteer editor and secretary. It is adjusted so as to cover only journals that have volunteer editors. For other journals, the cost of subject editors is included in the staff editing category.

Review

Input: Unreviewed manuscript.
Process: Review and critical annotation.
Output: Reviewed manuscript.

This category is part of the overall editing process, but reviewers work on a volunteer basis and are not paid by the publisher. The cost of this activity is thus isolated as another donated cost.

It is assumed that the reviewer is a scientist who has the same median salary as an author. The amount of time spent reviewing an article is estimated at 6 hours for manuscripts that are rejected and 6.25 hours for those that are accepted.

Typesetting

Input:　　　Edited manuscript.
Process:　Typesetting and graphics preparation.
Output:　　Final composed pages.

There are a number of methods of going from manuscript copy to a finished master image of a journal article ready for reproduction. Photocomposition is now frequently used, and this is the method assumed in the publication cost mode. The cost of composition is a function of the number of pages published, the makeup of those pages, and the cost of composing different types of pages. Average costs per page vary considerably depending on the composition method used. The average cost per page in 1977 was about $46. Comparable computer composition costs, derived from data obtained in a feasibility test of the Editorial Processing Center concept,[17] are about $36 per page.

Costs are first calculated on a per-journal basis depending on the composition method used. To determine total typesetting costs, per-journal costs are multipled by the number of journals. Since some of these costs are borne by foreign subscribers, the U.S. portion of the costs is calculated from the proportion of subscriptions that go to U.S. subscribers.

Subscription Runoff

Input:　　　Master image of a journal issue.
Process:　Reproduction of the master image, assembly into journal issue, distribution to subscribers.
Output:　　Distributed journal issue.

The category of subscription runoff covers printing and distribution of subscription copies and a number of other, somewhat miscellaneous functions such as promotion and production of microform. The print run for each issue of an average journal was calculated based on average subscriptions, and then the following model was used to determine printing costs:

$$Cp = M [C_1 + C_2 \times n + C_3 \times N + C_4 + C_5) \times n \times N]$$

with Cp = annual print cost, M = number of issues per year, n = number of pages per issue, and N = number of copies printed per issue, and with C_1 = the setup costs associated with one issue, C_2 = platemaking and collating costs per page, C_3 = binding and cover costs per copy, C_4 = labor and equipment costs for printing per impression (one copy of one page), and C_5 = paper costs per impression.

The second element in runoff costs is distribution, or postage and handling costs. These were computed by the publication cost model based on the number of subscriptions, the weight of the journal, and the type of postage required (second-class profit or nonprofit, foreign). Other runoff costs, such as maintaining the subscription list, were estimated at 15 percent of total runoff costs.

To develop total costs, printing, distribution, and other costs were summed and multiplied by the number of journals. As with prerun costs, a portion of runoff costs can be allocated to foreign subscribers, and only the U.S. portion of the runoff costs is included in the cost model.

Separates Runoff*

Input: Master image of a journal article.
Process: Reproduction of the master image.
Output: Reprint or preprint copies of the article.

The number of reprints distributed by authors was estimated in the author survey associated with the Statistical Indicators project.[18] More reprints are obtained by authors than are distributed, and publishers also distribute reprints directly to users. The estimate of the volume of reprints printed used in the cost model is twice the number of copies distributed by authors. This comes to 138 reprints per article over all

*Copies of journal articles are frequently requested from publishers. Fulfillment of these requests are called separates.

fields of science. The cost model for the reproduction of reprints is as follows:

$$C_r = A \times [C_1 + (C_2 \times N)]$$

With C_r = annual reprint cost, A = number of articles published, N = reprint copies per article, C_1 = set-up cost of a reprint run, and C_2 = print cost per reprint copy. The equation was used to arrive at total reprint costs per journal. It is assumed that all reprints are distributed within the United States. Distribution costs for the publisher are included in other runoff costs, and reprint distribution costs for the author are included in the category covering individual acquisition of reprints.

It is estimated that an average of five preprints are sent out for every published article. Reproduction of preprints is generally through photocopying, a process estimated to cost $0.21 per page in 1977, including labor, equipment, and supplies. The average length of a manuscript is twenty-four pages, or three times the number of typeset pages in the final article, so that the total 1977 cost per preprint runoff is $5.04. Costs for other years are similarly calculated. Costs associated with the distribution of preprints are included in the category covering individual acquisition of preprints.

Components of the Cost Model

Cost data maintained by publishers are generally incomplete because they omit donated costs not directly incurred by the publisher, and, more importantly, they often do not identify costs of specific activities. Most frequently, costs are associated with the various locations in which operations are performed. For this reason, we have constructed a model that calculates costs for particular operations and also can combine costs in various ways for comparisons with data from the accounting systems of various journals. The publication cost model is also set up so that the sensitivity of costs to changes in the publication process can be evaluated.

The cost model contains over a hundred parameters that define the journal characteristics, cost elements, and process variables involved. Cost elements and process variables were set at 1977 values, and most of these were held constant over the analyses. Exceptions were authors' and reviewers salaries and rejection rates, which vary among the fields of science. A number of journal characteristic parameters were also varied for each different analysis. Some of the most important journal

publication characteristics used in the cost model are given in
Table 4.21. All of the data were estimated from a sample of 377 journals.
Characteristics were observed in libraries for these sampled journals.

Journal publication characteristics vary substantially among fields of
science, so their average costs of publishing also vary. The journals in
Physical Sciences publish the greatest number of articles per journal
(207), whereas Social Sciences has the fewest (40). Since the number of
pages per article does not vary much, these numbers mean that the
number of article pages published per year is also greatest for journals
in Physical Science (1,365). Both Social Sciences (531) and Psychology
(436) are low. The average number of issues published reflects the
number of articles published per journal among fields. Averages for
journals in Other Sciences are 10.3 issues per year and in
Social Sciences 4.4 issues per year. One of the most important journal
characteristics affecting costs is the number of pages with graphics. The
proportion of graphic pages is very high in journals published in
Computer Science (53 percent) and Engineering (38 percent) and much
less in such fields as Psychology (8 percent) and Social Sciences
(5 percent). The difference between total pages published and article
pages published is accounted for by such items as book reviews, letters,
and advertising. For the most part the two figures are related. The
proportion of pages devoted to articles is greatest in Physical Sciences
(87 percent) and least in Other Sciences (63 percent). Kiloword pages
published is a way to normalize[19] so that one can better compare the
actual content among fields and over the years. Trends for publication
characteristics from 1960 to 1974 were presented in a King Research
report to NSF.[20]

The principal difference in journal characteristics among the three
sizes of journals is number of pages published. Leeson and Machlup
found[21] that in 1974 about three-quarters of the small publications
surveyed had fewer than 1,000 pages, while only half of the medium and
large journals had fewer than 1,000 pages. The average number of
pages per year for the three categories were 1,068, 1,752, and 1,764,
respectively. Our model utilizes similar proportions applied to the
average number of pages overall observed in 1977. Data on journal
characteristics other than number of pages were not available by size
of journal.

Prerun costs was the first major category estimated. The costs
shown in Table 4.22 are averages for all journals, a point of particular
note in the discussion of editing costs. Technical editing costs vary with
the number of articles published and, less significantly, with the number
of pages published. They also vary with the number of manuscripts
submitted. A journal that rejects many papers will, other things being

Table 4.21 Journal publication characteristics, 1977.

	Average Value of Characteristic							
	Issues per Year	Articles Published	Article Pages Published	Proportion of Graphics Pages	Total Pages Published	Kiloword Pages Published	Subscriptions	Subscription Price
Field of science								
Physical sciences	9.8	207	1,365	20	1,573	1,644	3,598	$64.55
Mathematics	6.4	81	829	26	986	701	4,358	37.63
Computer sciences	7.3	132	568	53	754	924	17,226	30.87
Environmental sciences	7.2	73	582	36	772	869	11,582	20.49
Engineering	6.9	99	576	38	790	896	8,483	35.57
Life sciences	7.7	97	582	16	849	789	6,620	25.13
Psychology	4.6	59	436	8	559	451	3,483	16.57
Social sciences	4.4	40	531	5	621	466	2,324	12.95
Other sciences	10.3	143	902	33	1,440	1,732	18,281	13.61
All fields	6.6	86	644	18	832	796	6,327	23.93
Size of journal								
Small	6.6	65	485	18	674	645	1,391	49.72
Medium	6.6	108	807	18	996	953	5,454	30.22
Large	6.6	108	807	18	996	953	27,450	17.21

Table 4.22   Prerun journal publication costs, 1977.

	All Fields
Editing	$29,000
Composition	38,100
Graphics	5,700
Total	72,800
Cost per kiloword page	91

Source: King Research, Inc. Journal Cost Model.

equal, show higher editing costs per published manuscript. In addition to these measurable differences, there are wide variations among journals in the amount of time devoted to editing. These variations appear to be related to subject matter, type of articles submitted, and editorial policy.

Composition costs refer to the cost of typesetting, often a purchased service. The costs used in the model primarily reflect the cost of purchased photocomposition, including different costs for standard text, subsidiary text, equations and symbols, and tables. These costs thus vary with the number of pages published and the complexity of the pages. Costs for figures and photographs are also related to the makeup of the published pages, with costs varying with the number and type of graphics.

If the values of journal prerun characteristics for small, medium, and large journals are held constant, except for the number of subscriptions, the costs can be used to illustrate an important difference related to size of journal. The prerun cost per subscriber comes to $52, $13, and $3 respectively for small, medium, and large journals. This dramatically indicates the plight of the small journal.

Table 4.23 gives estimates of printing, distribution, reprint, and other costs for all scientific and technical journals. Of these, printing and distribution are generally considered as runoff costs and reprints and other costs as miscellaneous costs. By far the greatest cost is for printing, usually a purchased service, which includes the cost of labor, equipment, paper, and other supplies.

Costs are dependent on both the numbers of pages and copies printed. For this reason, printing costs and total runoff costs are often presented in terms of cost per page per subscription or of cost per kiloword page per subscription. Runoff costs per page per subscription vary only a little in the range of 0.51 to 0.88 cents for the different fields of science. Put into terms of the volume of information provided, runoff costs per kiloword page per subscription range from 0.49 to 1.17 cents among the fields of science and .62 to 1.23 cents depending on the size of the journal.

Table 4.23   Runoff and miscellaneous journal publication costs, 1977.

	Costs
Runoff	
Printing	$27,700
Distribution	10,700
Miscellaneous	
Reprints	2,000
Other	7,100
Total	47,600
Runoff cost per page	
per subscription	0.73
Runoff cost per kiloword page	
per subscription	0.76

Source: King Research, Inc. Journal Cost Model.

Table 4.24   Journal publication costs: donated services, 1977.

	Costs
Author	$153,200
Subject editor	3,400
Reviewer	17,600
Subtotal (subject editor	
and reviewer)	21,100
Total	174,300

Source: King Research, Inc. Journal Cost Model.

Other costs include reprint costs and the residual category of costs not reported elsewhere, which might be promotion, advertising, distribution of reprints and back issues, or other similar activities. Together the total cost of runoff and miscellaneous activities averages $47,600 for all journals.

We have also estimated the total donated costs associated with authors, subject editors, and reviewers (Table 4.24). The author costs dominate, although the other ones are not insignificant.

Table 4.25 summarizes the cost model results for the entire publication process, giving prerun, runoff, other, and donated costs. Table 4.26 presents these by field of science. Average total costs (including the donated costs of subject editors and reviewers) are $141,500 for all scientific and technical journals. By size of journals, the average ranges from $88,300 for small journals to $308,100 for large journals.

The percentage breakdown of costs is very similar among the fields of science, in accordance with relatively small variation in the

Table 4.25 Total journal publication costs, 1977.

	All Fields
Prerun	$ 72,800
Runoff	38,400
Other	9,200
Total	120,400
Donated	21,100
Total, including donated	141,500
Total for all journals (millions)	629

Source: King Research, Inc. Journal Cost Model.

Table 4.26 Journal publication costs, by field of science, 1977

Field of Science	Prerun	Runoff	Other	Subtotal	Donated	Total	Total All Journals (millions)
Physical sciences	$152,400	$ 41,500	$12,800	$206,700	$46,100	$252,800	$ 70.3
Mathematics	88,200	30,000	7,200	125,400	22,100	$147,600	15.6
Computer sciences	84,300	96,500	20,275	201,100	31,200	232,300	15.8
Environmental sciences	69,200	65,700	13,800	148,700	17,900	166,600	39.0
Engineering	75,500	49,900	11,300	136,800	26,300	163,100	76.8
Life sciences	69,800	42,800	10,600	123,200	20,700	143,900	189.7
Psychology	48,900	16,500	4,500	70,000	16,500	86,500	19.7
Social sciences	49,700	12,700	3,200	65,600	11,100	76,700	109.5
Other sciences	110,100	155,700	31,200	297,100	45,000	342,100	108.1
All fields	72,800	38,400	9,200	120,400	21,100	141,500	629.3

Source: King Research, Inc. Journal Cost Model.

parameters for the different fields. However, the average number of subscriptions for the nine fields ranges from 2,324 to 18,281. The range of the average number of pages is 559 to 1,573. Prerun costs (including donated) account for about 67 percent or more of the total costs, runoff costs 20 to 30 percent, and other costs less than 10 percent.
Prerun costs thus dominate, a characteristic that is even more marked for small journals, where the cost of editing and preparing manuscripts for publication makes up a full 85 percent of total costs. For large journals, the prerun percentage drops to 30.

Among the fields of science, prerun costs account for a somewhat greater proportion (and runoff costs for less) in the Physical Sciences and Psychology, primarily because of the lower average number of subscribers in these fields. When donated costs are expressed as a proportion of total costs, the range is from 10 (Environmental Sciences) to 19 percent (Psychology) among the fields of science, and from 9 to 18 percent depending on the size of the journal.

Trends in Journal Publishing Costs

The changes that can be anticipated in journal publishing costs over the coming years are of critical interest. They could be drastic if the nature of the publication system were to change radically, but they will be significant even if the system remains essentially the same. Only projections of the current system will be discussed here.

Leeson and Machlup's data[22] provide some cost trends for publication up to 1975, reflecting actual costs obtained from as many as 108 scientific and technical journals.* Different journals reported data for different sets of years. The method used to produce consistent estimates was to begin with the data for 1974 (for the maximum number of journals) and compute a corresponding 1973 figure using the rate of change for all journals reporting both 1973 and 1974 data. In the same manner costs were estimated back to 1966 and forward to 1975. The results are shown in Table 4.27.

Per-page costs increased from $62 in 1966 to $119 in 1975, about 90 percent over the nine-year span. In constant dollar terms, the increase was only 16 percent, from $81 to almost $94. Over the same period, circulation also increased somewhat, so that the average cost per page per subscription shows a current dollar increase of 80 percent and a constant dollar increase of only 8 percent. Thus the 1966–1975 increases in publication costs for the average scientific and technical journal appear reasonable for the amount of inflation and increased production.

The differences in cost patterns among different groups of journals may stem from a number of factors. Declining circulation implies that prerun costs must be spread over a smaller group of subscribers, and this will have a detrimental effect on per subscriber costs. Outside a narrow range, changes in the number of pages or articles will noticeably affect costs by changing the base over which fixed costs are spread. The costs for a given group of journals may be modified because of changes in the length of articles, the amount of graphics, the number of advertising pages, the rejection rate for articles, or other characteristics.

*Special computer runs were made of Leeson and Machlup data for this study.

Table 4.27 Publication cost trends, 1966–1975.

Year	Average Cost per Page		Average Cost per Page per Subscription	
	Current Dollars	Constant[a] Dollars	Current Cents	Constant[a] Cents
1966	$62.20	$81.00	$1.56	$2.03
1967	69.00	87.30	1.67	2.11
1968	78.90	95.60	1.83	2.22
1969	81.80	94.30	1.85	2.13
1970	86.50	94.70	1.92	2.10
1971	88.50	92.20	2.00	2.08
1972	95.00	95.00	2.16	2.16
1973	97.40	92.10	2.20	2.08
1974	107.00	91.90	2.49	2.14
1975	119.10	93.60	2.80	2.20

Source: D. W. King and N. K. Roderer, *Systems Analysis of Scientific and Technical Communication in the United States,* Annex 2: *The Current Practices* (Rockville, Maryland: King Research, 1978), p. 111.
[a]1972 constant dollars were computed using the GNP implicit price deflator.

Finally procedures may be changed for a journal or group of journals (perhaps simplified or automated), and corresponding changes in costs will occur.

Our publication cost model takes into account most of these factors and therefore can be used to estimate costs under widely varying conditions. It has been used to estimate 1980 and 1985 costs for all science and technology journals, for journals in the nine fields of science, and for small, medium, and large journals. These estimates were based on application of the model using 1980 and 1985 projections for the values of over a hundred relevant parameters, including journal characteristics and process and cost factors.

Major costs involved in publishing include editing labor, typesetting, printing labor, paper, and postage. Editing, typesetting, and printing together make up the greatest part of all publishing costs, with each accounting for one-quarter or more of total publishing costs in 1975. Paper and postage each accounted for about 10 percent of total costs in 1975; they also represent areas where larger than average increases may occur in the future.

Editing is a highly labor-intensive activity, involving the efforts of managing and subject editors, copy editors, support personnel, and reviewers. Average hourly salaries between 1960 and 1985 for each of these groups are indicated in Table 4.28. The 1980 and 1985 data are

Table 4.28 Salaries of editors, secretaries, and reviewers, 1960–1985.

Year	Subject Editor Hourly Salary	Subject Editor Five-Year Change	Copy Editor Hourly Salary	Copy Editor Five-Year Change	Secretary Hourly Salary	Secretary Five-Year Change	Reviewer Hourly Salary	Reviewer Five-Year Change
Current dollars								
1960	$5.04		$3.03		$2.38		$3.77	
1965	6.07	20%	3.57	18%	2.72	14%	4.57	21%
1970	7.76	28	4.56	28	3.41	25	6.05	32
1975	10.67	38	6.28	38	4.52	33	8.33	38
1980	14.72	38	8.56	36	6.38	41	11.44	37
1985	16.97	15	9.83	15	7.27	14	14.52	27
Constant dollars								
1960	$7.34		$4.41		$3.47		$5.49	
1965	8.17	11	4.80	9	3.66	5	6.15	12
1970	8.49	4	4.99	4	3.73	2	6.62	8
1975	8.39	–1	4.94	–1	3.55	–5	6.55	–1
1980	8.93	6	5.19	5	3.87	9	6.93	6
1985	9.43	6	5.46	5	4.04	4	8.07	16

Table 4.29   Per page typesetting costs of basic text, 1960–1985.

Year	Current Dollars		Constant Dollars	
	Per Page Cost	Five-Year Change	Per Page Cost	Five-Year Change
1960	$19.00		$27.67	
1965	21.80	15%	29.33	6%
1970	28.20	29	30.87	5
1975	42.40	50	33.34	8
1980	57.00	34	33.87	2
1985	66.70	17	31.78	–6

Source: D. W. King and N. K. Roderer, *Systems Analysis of Scientific and Technical Communication in the United States,* Annex 2: *The Current Practices* (Rockville, Maryland: King Research, 1978), p. 114.

Table 4.30   Printing cost elements, 1960–1985.

Year	Current Dollars		Constant Dollars	
	Wage Index for Union Printers	Five-Year Change	Wage Index for Union Printers	Five-Year Change
1960	$3.23		$4.70	
1965	3.73	15%	5.01	7%
1970	5.47	47	5.99	19
1975	7.80	43	6.13	3
1980	10.20	31	6.06	–1
1985	12.50	23	5.97	–1

Source: D. W. King and N. K. Roderer, *Systems Analysis of Scientific and Technical Communication in the United States,* Annex 2: *The Current Practices* (Rockville, Maryland: King Research, 1978), p. 115.

projections based on more complete 1960–1975 data. The greatest current dollar increases are indicated for the 1970–1975 period, with 1975–1980 increases almost as large.

Typesetting costs depend on the type of material being set (as well as on the technology used). The costs shown in Table 4.29 are for a standard text page, reflecting the observed and anticipated rates of growth for all typesetting. The table is based only on photocomposition costs and not the costs of new technology expected to be utilized increasingly. Increases are generally similar to those for salaries. Constant dollar price increases are reasonably consistent at about 6 percent over each five-year period.

Printing costs are a composite of labor, equipment and supply, and paper costs. Table 4.30 shows a wage index for union printers. Constant

Table 4.31   Paper cost elements, 1960–1985.

	Current Dollars		Constant Dollars	
Year	Paper Costs (per Piece)	Five-Year Change	Paper Costs (per Piece)	Five-Year Change
1960	$0.210		$0.31	
1965	0.210		0.28	−10%
1970	0.270	29	0.30	7
1975	0.302	12	0.24	−20
1980	0.342	13	0.20	−17
1985	0.378	11	0.18	−10

Source: D. W. King and N. K. Roderer, *Systems Analysis of Scientific and Technical Communication in the United States,* Annex 2: *The Current Practices* (Rockville, Maryland: King Research, 1978), p. 115.

dollar increases were fairly large in the 1965–1970 period and are expected to reach a significant level again in the 1980–1985 period.

Paper costs have been the subject of considerable attention in the journal publishing industry in recent years, with apprehension voiced concerning erratic changes in costs and the possibility of significant rises. For a closely related industry, newspapers, paper prices have increased steeply over the last few years. For book paper, however, increases have not been nearly as large. Table 4.31 shows paper costs per piece (two pages) over the years. Between 1975 and 1985, costs are expected to increase about 2 percent annually in current dollar terms and to decrease about 3 percent in constant dollar terms.

The final major cost element to be considered is postage. The model includes a variety of postage rates, including those for mailing manuscripts and proofs between author and publisher and various rates for mailing journal issues. The cost of mailing a journal issue depends on where the issue is going, what portion of the issue is devoted to advertising, and whether the publication is nonprofit. Rates shown in Table 4.32 are for periodicals sent within the United States and containing less than 10 percent advertising. Separate figures are given for nonprofit and other periodicals. The increases projected are probably conservative; they are, nonetheless, significant in the 1980–1985 period.

Cost Projections

The major cost factors in journal publishing, together with projections of journal characteristics and process factors, were put into

Table 4.32   Second-class postal rate per pound, 1960–1985.

	Nonprofit Periodicals		Other Periodicals	
Year	Rate	Five-Year Change	Rate	Five-Year Change
Current dollars				
1960	$0.015		$0.023	
1965	0.018	20%	0.028	22%
1970	0.021	17	0.034	21
1975	0.030	43	0.056	65
1980	0.034	13	0.064	14
1985	0.043	26	0.081	27
Constant dollars				
1960	0.022		0.033	
1965	0.024	9	0.038	15
1970	0.023	−4	0.037	−3
1975	0.024	4	0.044	19
1980	0.021	−12	0.039	−11
1985	0.024	14	0.045	15

Source: D. W. King and N. K. Roderer, *Systems Analysis of Scientific and Technical Communication in the United States,* Annex 2: *The Current Practices* (Rockville, Maryland: King Research, 1978), p. 116.
Note: Excluding per-piece charge when applicable; for journals with less than 10 percent advertising.

our publication cost model. Total publishing costs for 1980 and 1985 were thus derived (Table 4.33). The cost projections are $171,900 for 1980 and $202,900 for 1985. The eight-year change from 1977 to 1985 is estimated to be 90 percent, or an average of about 8 percent annually.

Summary of Journal Publishing Costs

Some costs have been excluded from the totals because they were associated with foreign subscribers. Publishing costs presented up to this point have not reflected these exclusions. Table 4.34 gives journal publication costs included in the total journal system cost calculations. These range from $15 million in Computer Sciences to $185 million in Life Sciences. The total is $618 million.

FEDERAL CONTRIBUTION TO PUBLISHING

One aspect of the federal contribution to publication support is the direct publication of journals by various federal agencies. The

Table 4.33 Total journal publication costs, 1980 and 1985.

Cost Category	1980	1985
Prerun	$88,100	$98,200
Runoff	46,800	57,800
Other	11,200	13,700
Subtotal	146,000	169,700
Donated	25,800	33,200
Total	171,900	202,900
Total all journals (millions)	817.0	1,076.2

Source: King Research, Inc. Journal Cost Model.

Table 4.34 Journal publication costs allocated to U.S. subscribers, by field of science and size, 1977.

	Publication Costs (millions)
Field of science	
Physical sciences	$67.8
Mathematics	15.5
Computer sciences	15.1
Environmental sciences	40.7
Engineering	75.1
Life sciences	185.3
Psychology	18.4
Social sciences	108.3
Other sciences	110.8
Total	618.0
Size of journal	
Small	191.7
Medium	248.7
Large	182.9

Source: King Research, Inc. Journal Cost Model.

Government Printing Office's list of federal serials includes 149 scientific and technical journals, distributed among the fields of science as shown in Table 4.35. Also indicated in the table is the estimated publishing cost per journal, and the resulting federal contribution for each field. The total federal contribution from the publication of federal journals is estimated at approximately $21 million, 3.4 percent of total U.S. scientific and technical journal publication costs.

The federal government also contributes to publishing through its support of page charges and subscriptions by federal employees and libraries and by federally funded individuals and libraries. In the

Table 4.35 Federal support of journal publications, by field of science, 1977.

Field of Science	Number of Journals Federally Published	Per Journal Cost (millions)	Total Federal Contribution (millions)
Physical sciences		$0.24	
Mathematics	3	0.15	$0.4
Computer sciences		0.22	
Environmental sciences	32	0.17	5.4
Engineering	35	0.16	5.6
Life sciences	26	0.14	3.6
Psychology		0.08	
Social sciences	47	0.08	3.7
Other sciences	6	0.35	2.1
All fields	149		20.8

Source: King Research, Inc. Journal Cost Model.

user survey, scientists indicated that only 0.7 percent of their individual subscriptions were paid by the federal government, for about $2.9 million in federal support. For library subscriptions, we limit our estimates to the U.S. journal expenditures of the federal scientific and technical libraries, roughly $4.0 million. These data are reflected in the summary of federal contributions given in Chapter 8.

REFERENCES

1. D. W. King, D. D. McDonald, N. K. Roderer, C. G. Schell, C. G. Schueller, and B. L. Wood, *Statistical Indicators of Scientific and Technical Communication (1960–1980),* 2d ed. (Rockville, Maryland: King Research, 1977), pp. 70–73.
2. British Library Lending Division (BLLD) holdings.
3. B. M. Fry and H. S. White, *Publishers and Libraries: A Study of Scholarly and Research Journals* (Lexington, Mass.: Lexington Books, 1976).
4. *Ulrich's International Periodicals Directory* (New York: Bowker, 1975-1976).
5. D. S. Price, *Little Science Big Science* (New York: Columbia University Press, 1963).
6. C. Herring, "A Study of Primary Journal Economics," In *Report of the Task Group on the Economics of Primary Publication,* Committee on Scientific and Technical Communication (SATCOM),

National Academy of Sciences-National Academy of Engineering (Washington, D.C.: National Academy of Sciences, 1970).

7. International Council of Scientific Unions Abstracting Board, *Some Characteristics of Primary Periodicals in the Domain of the Physical Sciences* (Paris: UNESCO, 1966).

8. *Primary Scientific Publications,* report prepared for the ICSU/UNESCO Joint Study on the Communication of Scientific Information and on the Feasibility of a Worldwide Science Information System (Chairman and rapporteur: Professor G. A. Boutry) (Paris: UNESCO, October 1967).

9. National Science Foundation, *Characteristics of Scientific Journals, 1949–1959* (Washington, D.C.: Government Printing Office, 1964).

10. *Ulrich's International Periodicals Directory.*

11. Ibid.

12. Ibid.

13. Herring, "Study of Primary Journal Economics."

14. King, McDonald, Roderer, Schell, Schueller, and Wood, *Statistical Indicators.*

15. National Science Foundation, *Characteristics of Scientific Journals.*

16. K. W. Leeson and F. Machlup, "A Catalog of Statistical Data on the Economic Characteristics of Primary Research Journals in the Sciences and Technology," Preliminary Findings, March 18, 1977.

17. W. Burke, *Final Report on Editorial Processing Center Operational Experiments* (Rockville, Maryland: Aspen Systems Corporation, 1977).

18. King, McDonald, Roderer, Schell, Schueller, and Wood, *Statistical Indicators.*

19. Herring, "Study of Primary Journal Economics."

20. King, McDonald, Roderer, Schell, Schueller, and Wood, *Statistical Indicators.*

21. Leeson and Machlup, "Catalog of Statistical Data."

22. Ibid.

Chapter 5

Libraries and Secondary Services

Libraries and secondary services are important intermediaries between publishers and users of scientific and technical information. They play major roles in the acquisition and storage of the literature, its organization and control, and its identification and access by users. In carrying out the last four activities, libraries rely on secondary services to provide abstract and index publications and computer bibliographic searching. These functions are particularly critical in light of the large amount of literature generated and the long time over which it may be used.

LIBRARY ACTIVITIES

Number of Libraries in the United States

Libraries are generally classified as academic, public, special, or school according to their primary clientele. In general, academic and special libraries play a more significant role in the dissemination of scientific and technical information than do public and school libraries. Two groups of libraries that fit into more than one of the four classes and serve as important sources of scientific and technical journals are research libraries and federal libraries.

Research libraries, such as the membership of the Association of Research Libraries (ARL), represented by large libraries, build and maintain extensive collections of research literature recording the achievements of science and technology. These libraries assume the major responsibility for ensuring that research materials are available to scientists and other scholars. A large portion of journal authors come from universities. Without research libraries, the scientists would be impeded in their scientific investigation since most would have no obvious source for a record of past scientific achievement or for a wide selection of current scientific publications. The existence in the

129

United States of many large and important research libraries tends to ensure that any scientific document of value is and will continue to be available; serious gaps in coverage are not allowed to occur.

The federal libraries include a number of specialized libraries. Of the 2,313 federal libraries identified in 1972, 43 percent were classified as special or technical. Among these were the three national libraries—the Library of Congress, the National Library of Medicine, and the National Agricultural Library—each of which provides extensive informational services to both individual scientists and to other libraries.

Figure 5.1 shows the total number of academic, special, and public libraries from the year 1960. The number of academic libraries grew considerably in the 1960s but leveled off in the 1970s and is projected to grow only slowly in the future. Special libraries have increased significantly in number over the years, and we project that there will be 12,750 special libraries in 1985. The number of public libraries is increasing more slowly. Considering the three types of libraries together, in 1977 there were 2,839 academic libraries, 11,000 special libraries, and 9,133 public libraries, for a total of 22,972. These are, of course, neither equivalent in size nor equally important to the communication of scientific and technical literature.

Libraries acquire a selection of journals, as they do other types of materials, appropriate to their clientele. Academic and research libraries tend to maintain large retrospective collections of journals, consistent with their role as archival sources, as well as performing the normal library functions. Special libraries may place considerable importance on the journal literature as a very current source of information.

Information centers, closely related to libraries, also play a role in the scientific and technical communication process. Like special libraries, these centers are characterized by serving in limited subject areas; in addition, the term *information center* implies a greater depth of analysis and control and frequently more advanced services, such as evaluation and synthesis of information.[1] Information centers may also include raw data among their holdings.

The concept of an information center is sometimes more broadly interpreted to encompass any unit involved in the storage, manipulation, or provision of information. This would include information analysis centers, data banks, and the like. Our use of the term, however, is restricted to the narrower definition, and other information service organizations are discussed under the specific functions they serve.

Data on libraries and information centers are available from a number of sources, the most wide ranging of which is the Library Surveys Branch of the National Center for Education Statistics (NCES)

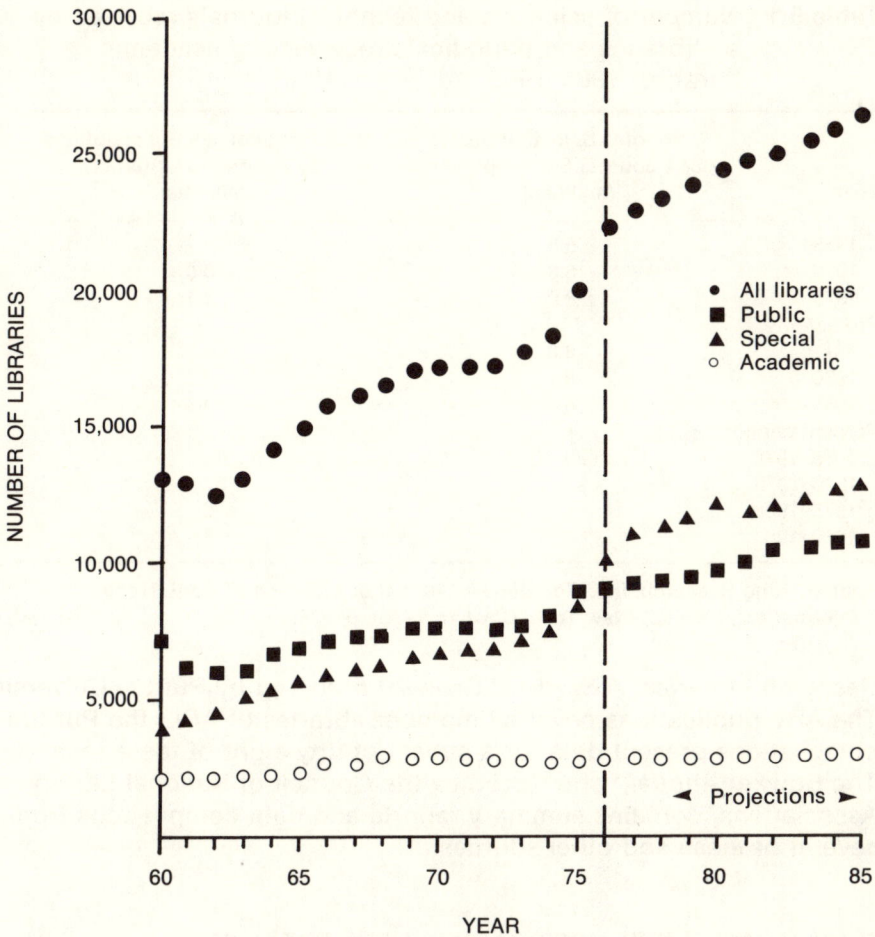

Figure 5.1 Growth of U.S. libraries by type, 1960–1985.

of the Office of Education. The Office of Education has collected and published library statistics with some regularity since the 1870s, and work currently in process covers public, school, academic, and special libraries, as well as library manpower. Under development is the Library General Information Survey (LIBGIS) program, intended to produce a national library statistics data system that will provide consistent and comprehensive data on all libraries.

Other important sources of library data, all annual, are the *American Library Directory,*[2] *American Library Statistics*[3] and *ARL Statistics*[4] prepared by the Association of Research Libraries (ARL), and *Academic*

Table 5.1   Number of scientific and technical journals received by
all libraries and periodicals received, by academic
libraries, 1965–1985.

Year	No. of U.S. Institutional S&T Journal Subscriptions (millions)	No. of Periodicals Subscriptions by Academic Libraries (millions)
1965	5.5	2.7
1970	6.6	3.0
1975	8.0	4.1
Projections		
1977	8.0	4.2
1980	8.1	4.3
1985	8.6	4.5
Percent change		
1965–1970	20	11
1970–1975	21	37
1975–1980	1	5
1980–1985	6	5

Source: King Research, Inc.; *The Bowker Annual of Library and Book Trade
Information,* 1976 ed. (New York: Bowker, 1976), p. 236.

Research Libraries: A Study of Growth[5] prepared by Purdue University.
The ARL publications cover all member libraries of ARL; the Purdue
compilations present data for a subset of fifty-eight of these institutions.
The *Bowker Annual,*[6] sponsored by the Council of National Library
Associations, contains summary reports and data compilations from
several of these and other sources.

Number of Scientific and Technical Journals Acquired

The best indication of the volume of U.S. scientific and technical
journals acquired by libraries is the estimate of library subscriptions
provided in Chapter 4 and repeated, for the years 1965 to 1975, in
Table 5.1. These journals are purchased by all types of libraries.
Libraries also acquire foreign materials, nonjournal periodicals, and, of
course, nonscientific materials. The second column of Table 5.1
indicates the total number of periodicals (scientific and technical and
otherwise, journals and nonjournals, U.S. and foreign) received by
academic libraries only. Both sets of data suggest that library
acquisitions are now holding steady rather than increasing, and small
increases are projected for the future.

Library Expenditures

Because total library expenditures determine to some extent the amount that will be spent on journals, the growth of these items is of great interest in the current context. Table 5.2 gives the total expenditures for each of the groups of libraries and the scientific and technical portion of these expenditures for these groups. No figures are available for special libraries, so data for a subset of these (federal scientific and technical libraries) are presented. Data for the ARL libraries are given in terms of a constant base of fifty-eight libraries.

The figures show generally larger increases in the 1960s than in the early 1970s. Data for 1975, however, show substantial 1974–1975 increases of 10 percent for academic libraries and 11 percent for public libraries. Because of these increases, projections to 1985 have been recomputed using population served as the independent variable. They suggest increases of about 7 percent annually to 1985, as compared to earlier projections of about 6 percent.

In order to construct a total expenditure figure for scientific and technical library activities, it is necessary to determine for each library type what part of the total expenditures is devoted to science and technology. The major factors in this determination are the proportion of material expenditures and the proportion of service expenditures (cataloging, circulation, reference, interlibrary loan) for science and technology. These figures are not generally available, but a number of partial indicators (mainly for individual libraries) suggest that around 50 percent of academic library expenditures and 10 percent of public library expenditures are related to scientific and technical information. Adding these figures to the total expenditures for federal scientific and technical libraries yields the total expenditures shown in Table 5.2. These figures exclude nonfederal special libraries* and thus underestimate total expenditures, perhaps by as much as 20 percent.

The growth of total expenditures is plotted in current and constant dollars in Figure 5.2. Total expenditures have increased considerably in recent years, with a 1974–1975 increase of 10 percent. In constant dollars, this increase was 1 percent. Projections indicate that library expenditures will rise to $1,189 million in current dollars and $693 million in constant dollars by 1980. Earlier projections were for $1,057 and $485 million dollars.

One item of particular interest is the average library expenditure per scientist or engineer. Since a major portion of library expenditures is by

*Research libraries fall into the categories of academic and special libraries.

Table 5.2 Total and scientific and technical expenditures for academic, public, and federal scientific and technical libraries, 1960–1980 (in millions).

Year	All Expenditures			Total Expenditures	All Expenditures[b]
	Academic Libraries	Public Libraries	Federal S&T Libraries		
1960	$159	$397	$27[a]	$146	$55
1961	184	415	32[a]	166	62
1962	213	443	37[a]	188	69
1963	247	470	43[a]	214	77
1964	276	517	48[a]	238	87
1965	320	554	53	268	101
1966	366	600	64[a]	307	113
1967	416	692	76[a]	353	134
1968	510	766	88[a]	420	150
1969	585	830	99	475	166
1970	650	923	103	520	190
1971	737	1,024	117	588	199
1972	796	1,079	133	639	205
1973	867	1,227	136	692	222
1974	960	1,301	153[a]	763	241
1975	1,056	1,440	170[a]	842	265
Projections					
1977	1,232	1,687	193	977	295
1980	1,504	2,084	299	1,259	338
1985	2,030	2,918	374	1,681	406
Percent change					
1960–1965	101	40	96	84	84
1965–1970	103	67	94	94	88
1970–1975	62	56	65	62	39
1975–1980	42	45	35	150	28
1980–1985	35	40	25	34	20

Sources: *The Bowker Annual of Library and Book Trade Information,* 4th–21st eds. (New York: Bowker, 1961–1976); H. Goldhor, "Indices of American Public Library Statistics," Graduate School of Library Science, University of Illinois, 1960–1975; P. N. Vlannes, J. W. Hodges, L. J. Holland, K. Kay, D. R. Mitchell, A. D. Searle, and D. J. Shearin, *Report of the Ad Hoc Group: Federal Agency Obligations for Management, Processing and Transfer of Scientific and Technical Information, Data and Technology (FY 1969–1973)* (Washington, D.C.: Federal Council for Science and Technology, Committee for Scientific and Technical Information, September 1972); O. C. Dunn, D. L. Tolliver, and R. S. Tolliver, *The Past and Likely Future of 58 Research Libraries 1951–1980: A Statistical Study of Growth and Change,* 1970–1971 ed. (West Lafayette, Indiana: Purdue University, 1972): Association of Research Libraries, *American Library Statistics* (Washington, D.C.: Association of Research Libraries, 1974–1976).
[a]Estimates.
[b]For fifty-eight research libraries.

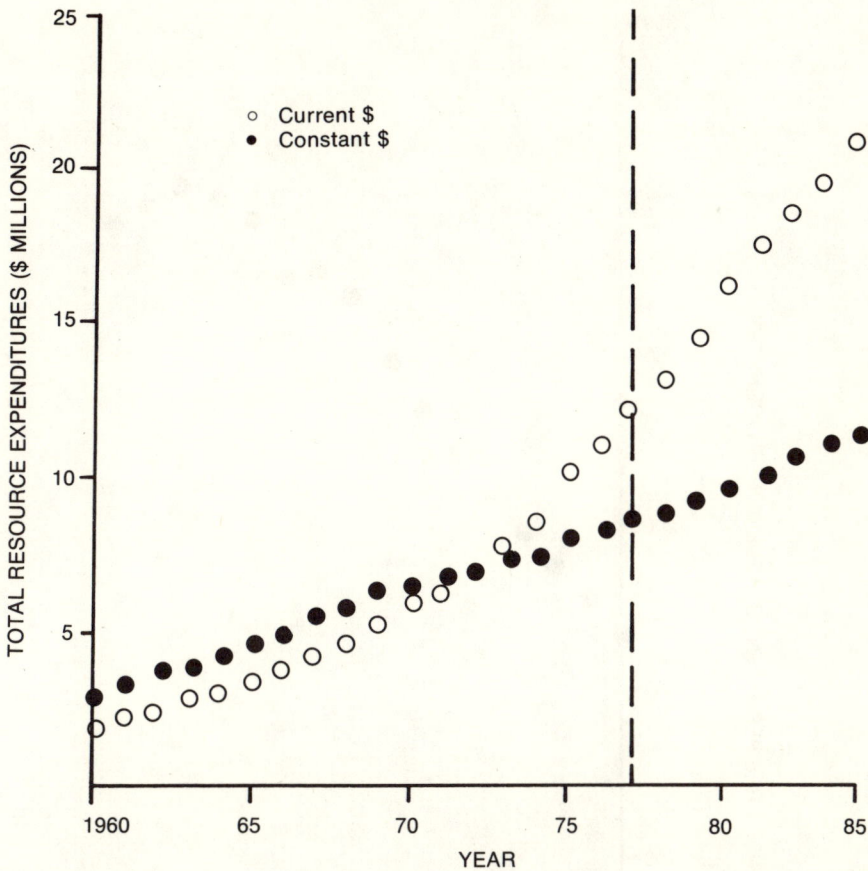

Figure 5.2 Growth of total resource expenditures, 1960–1985.

academic libraries, science and technology graduate students as well as scientists are included in the estimates of scientific manpower. Average library expenditure per scientist, using these manpower figures and constant dollar expenditures, is shown in Figure 5.3. It ranges from $166 in 1960 to $301 in 1973 and then drops slowly over the next few years. The average increased only 31 percent between 1965 and 1975, in contrast with the 83 percent increase in total constant dollar expenditures.

Library Materials Processing

Library processing activities include selection and acquisition of materials, cataloging, and storage and maintenance of the collection.

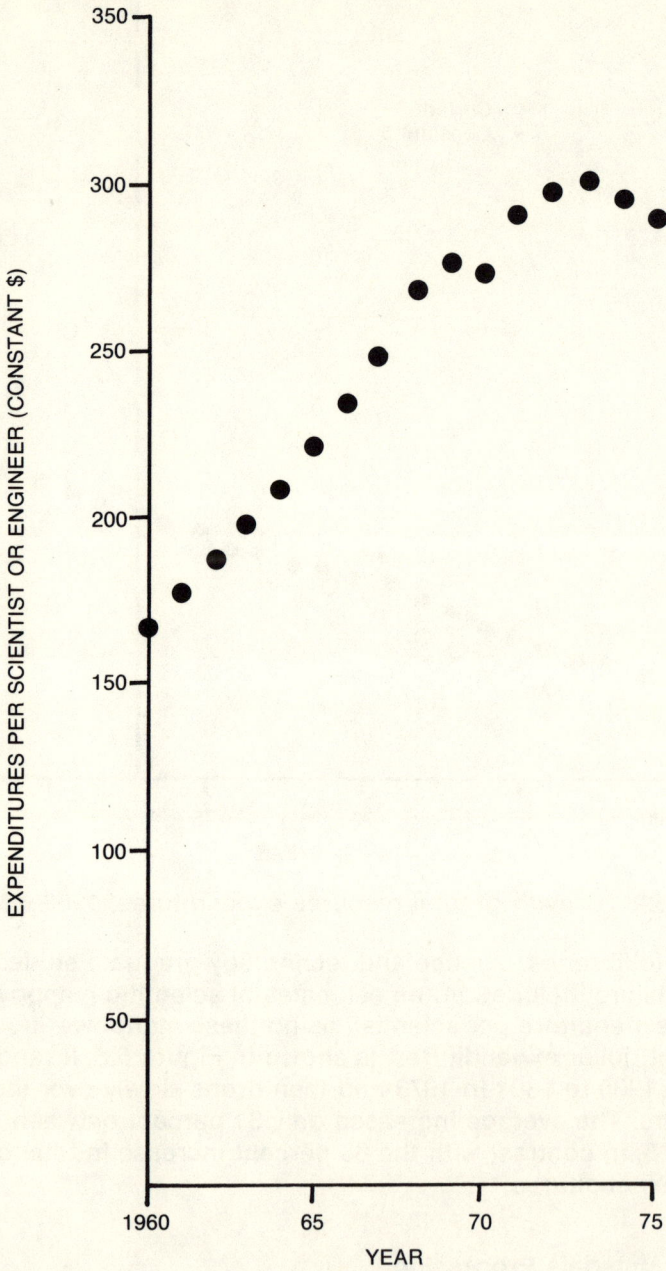

Figure 5.3 Library scientific and technical expenditure per scientist and engineer, 1960–1975.

Table 5.3   Volumes added to academic libraries, 1960–1974.

Year	Academic Libraries		Fifty-eight Research Libraries	
	Volumes Added	Change (%)	Volumes Added	Change (%)
1960	9,400		2,843	
1961	10,900	16	3,045	7
1962	12,300	13	3,369	11
1963	13,600	11	3,845	14
1964	15,000	10	4,054	5
1965	18,000	20	4,522	12
1966	20,000	11	4,992	10
1967	22,000	10	5,430	9
1968	25,000	14	6,420	18
1969	26,000	4	5,872	−9
1970	26,000		6,207	6
1971	25,000	−4	6,193	
1972	25,000		6,800	10
1973	25,000		6,200	−9
1974	25,000		5,700	−8
1975	25,000		5,500	−3

Sources: *The Bowker Annual of Library and Book Trade Information,* 4th–21st eds. (New York: Bowker, 1961–1976); O. C. Dunn, B. L. Tolliver, and R. S. Tolliver, *The Past and Likely Future of 58 Research Libraries, 1951–1980: A Statistical Study of Growth and Change,* 1970–1971 ed. (West Lafayette, Indiana: Purdue University, 1972); Association of Research Libraries, *American Library Statistics* (Washington, D.C.: Association of Research Libraries, 1960–1972).

Costs associated with these activities are the actual cost of the materials acquired and also a larger cost for the staff time involved. Important factors are the volume and type of materials added and the methods used in processing.

Table 5.3 shows the number of volumes added to academic libraries and the fifty-eight research libraries over the years. In the academic libraries, the number of volumes added annually has remained constant since 1971, at about 6 percent of their total holdings shown in Table 5.4. These figures show that the net increase in holdings in libraries (the difference between their total holdings from one year to the next) is much less. Between 1974 and 1975, for example, holdings increased by 10 million volumes as compared with acquisitions of 25 million. This means that 15 million volumes were withdrawn, for a discard rate of about 4 percent. The net increase, then, is 2 percent.

In the fifty-eight research libraries, acquisitions decreased, substantially in 1973 and 1974. The decrease was not as great in 1975.

Table 5.4 Holdings of academic libraries, 1960–1974.

Year	Academic Libraries		Fifty-eight Research Libraries	
	Holdings (000)	Change (%)	Holdings (000)	Change (%)
1960	189		73	
1961	201	6	76	4
1962	215	7	81	6
1963	227	6	84	4
1964	240	6	86	2
1965	265	10	91	6
1966	283	7	96	5
1967	295	4	101	5
1968	305	3	114	13
1969	329	8	110	–4
1970	354	8	116	5
1971	380	7	121	4
1972	405	7	126[a]	4
1973	407	0.5	131[a]	4
1974	426	5	135[a]	3
1975	436	2	142[a]	5
Percent changes				
1960–1965		40		25
1965–1970		34		27
1970–1975		23		22

Sources: *The Bowker Annual of Library and Book Trade Information,* 4th–21st eds. (New York: Bowker, 1961–1976); O. C. Dunn, B. L. Tolliver, and R. S. Tolliver, *The Past and Likely Future of 58 Research Libraries, 1951–1980: A Statistical Study of Growth and Change,* 1970–1971 ed. (West Lafayette, Indiana: Purdue University, 1972); Association of Research Libraries, *American Library Statistics* (Washington, D.C.: Association of Research Libraries, 1960–1972).
[a]Estimates.

Acquisitions in 1975 were about 4 percent of holdings, with the net change in holdings slightly larger than the acquisitions rate because new reporting units were added.

The term *volume* refers generally to books and bound periodicals. Also acquired by libraries are a large number of periodical or serial publications. The total titles received by academic libraries are shown in Table 5.5. Acquisition rates increased substantially in the late 1960s and early 1970s, with a leveling off since 1972. The average number of periodicals received in 1975 by an individual academic library was about 1,400.

The costs of both books and periodicals are included in the materials expenditures presented in Table 5.6. Figures are given for academic, public, and federal scientific and technical libraries and, from

Table 5.5 Number of periodicals received by academic libraries, 1964–1975.

Year	No. of Periodicals Received (Millions)
1964[a]	2.0
1965[b]	2.7
1966[b]	3.0
1967[b]	3.4
1968[b]	2.5
1969	2.5
1970	3.0
1971	3.6
1972	3.9
1973	3.9
1974	4.0
1975	4.1
Percent change	
1965–1970	11
1970–1975	37

Source: *The Bowker Annual of Library and Book Trade Information,* 4th–21st eds. (New York: Bowker, 1961–1976).
[a]Figures not available for 1960–1963.
[b]Figure is for serials, which includes periodicals, annuals, proceedings, transactions, and similar materials.

these, an estimate of funds expended on scientific and technical books and periodicals is derived. Data are also given for the fifty-eight research libraries.

Materials expenditures have increased in recent years, but more slowly than total expenditures. For example, the 1974–1975 increases for all scientific and technical libraries were 10 percent for total expenditures and only 4 percent for material expenditures. This indicates an increase of 13 percent in nonmaterials cost, predominantly labor. Figures for academic libraries—a 2 percent increase in materials and a 10 percent increase in total costs—are particularly noteworthy.

Nonmaterials expenditures include costs associated with both processing of materials and with services performed for library users. The two are difficult to separate, but there are indications that as much as two-thirds of nonmaterial costs are for processing. For all scientific and technical libraries in 1977, this figure would be $492 million, or more than twice the actual materials expenditures. Increases observed in nonmaterials cost are substantial.

Increasingly many of the activities involved in materials processing are becoming automated or computer assisted. This applies particularly

Table 5.6 Materials expenditures for academic, public, and federal scientific and technical libraries, 1960–1980 (in millions).

Year	All Materials			Total S&T Materials Expenditures	Materials Expenditures, Fifty-eight Research Libraries
	Academic Libraries	Public Libraries	Federal S&T Libraries		
1960	$48	$64	$6[a]	$36	$17.9
1961	56	66	7[a]	41	20.6
1962	65	71	8[a]	48	23.1
1963	81	71	10[a]	58	26.0
1964	90	72	11[a]	63	30.0
1965	111	72	12	75	36.3
1966	134	90	15[a]	91	39.0
1967	156	104	17[a]	105	47.9
1968	188	115	20[a]	126	53.3
1969	213	141	23	144	57.8
1970	230	175	24	156	66.9
1971	247	134	27	169	67.3
1972	260	205	31	182	64.8
1973	282	209	31	193	71.9
1974	300	221	35[a]	207	78.7
1975	305	245	38[a]	215	83.7
Projections					
1977	334	290	43	239	86.5
1980	376	360	55	279	90.6
1985	444	522	75	349	96.9
Percent change					
1960–1965	131	12	100	108	103
1965–1970	107	143	100	108	84
1970–1975	33	40	58	38	25
1975–1980	23	47	45	30	8
1980–1985	18	45	36	25	7

Sources: *The Bowker Annual of Library and Book Trade Information,* 4th–21st eds. (New York: Bowker, 1961–1976); H. Goldhor, "Indices of American Public Library Statistics," Graduate School of Library Science, University of Illinois, 1960–1975; P. N. Vlannes, J. W. Hodges, L. K. Holland, K. Kay, D. R. Mitchell, A. D. Searle, and D. J. Shearin, *Report of the Ad Hoc Group: Federal Agency Obligations for Management, Processing and Transfer of Scientific and Technical Information, Data and Technology (FY 1969–1973)* (Washington, D.C.: Federal Council for Science and Technology, Committee for Scientific and Technical Information, September 1972; King Research, Inc.; O. C. Dunn, D. L. Tolliver, and R. S. Tolliver, *The Past and Likely Future of 58 Research Libraries, 1951–1980: A Statistical Study of Growth and Change,* 1970–1971 ed. (West Lafayette, Indiana: Purdue University, 1972); Association of Research Libraries, *American Library Statistics* (Washington, D.C.: Association of Research Libraries, 1974–1976).
[a]Estimates.

to cataloging but also to ordering, serial check-in, and other operations. Although the effect of this factor cannot yet be seen in the overall picture, evidence from individual libraries generally suggests that automation will be a critical element in slowing down the rate of increase in nonmaterial costs.

Among the variety of services provided by libraries to their patrons are reference, circulation, and interlibrary loan. Data on these activities are generally maintained by individual libraries but not reported to a central agency. The National Center for Education Statistics (NCES) does collect circulation and interlibrary loan data for some types of libraries and is beginning to collect data on reference activities.

Some new data on services in public libraries were released in 1977 by NCES and the University of Illinois. Reporting on 1974 circulation and interlibrary loan transactions, NCES gave figures of 893 million direct circulations and 2.4 million interlibrary loans provided by 8,307 public libraries. The circulation index compiled by the University of Illinois Library School, in turn, indicated a 10 percent increase in circulation from public libraries between 1974 and 1975 but only a 1 percent increase between 1975 and 1976 and none between 1976 and 1977.

Like other organizations involved in the communication of scientific and technical information, libraries in recent years have been faced with substantial increases in the volume of materials available, in the demands made for their services, and in costs. Chief among the measures taken to deal with these pressures have been the automation of library processes and the formation of resource-sharing networks. In many instances, automation has served as the driving force for formation of a new network.

A number of new networks came into being in the mid-1970s, including several that became participants in the Ohio College Library Center (OCLC). This organization, the major computerized cataloging service today, allows a member library access to over 2 million cataloging records created by the Library of Congress and by other participating libraries. In 1976–1977, more than 7.8 million books were cataloged using the OCLC system, an increase of 36 percent over the previous year's figure.

Among the major U.S. libraries are the Library of Congress (LC), the National Library of Medicine (NLM), and the National Agricultural Library (NAL). Like other libraries, they are responsible for collecting, preserving, and providing access to the nation's literature. Each also plays a leadership role in its respective specialized library community. LC is the largest national library in the world, with collections totaling over 16 million volumes and pamphlets, 31 million pieces of manuscript, 8 million photographic negatives and slides, 3 million maps, and 3 million music volumes and pieces. It performs a range of services for

Congress and for other libraries, among them the provision of cataloging information in both printed and machine-readable form. Its printed cards have been sold to libraries since 1901, and its computer tapes containing cataloging records have been distributed since 1968. The latter service, called the MARC Distribution Service, has provided a major impetus for automation in libraries. Among the functions performed by the NLM is the coordination of the Regional Medical Library Program and the provision of extensive computer-based bibliographic services. The NAL operates as the national resource library for agriculture and related fields, and also has a computer-based bibliographic data system. According to a 1972 survey of federal libraries,[7] the three national libraries together account for over 70 percent of the expenditures for federal scientific and technical libraries. In 1972 the budgets of the three libraries were $74, $17, and $4 million, respectively, for a total of $95 million.

Impact of Interlibrary Loans on Scientific and Technical Journals

Library networking and resource sharing are increasing, involving more extensive interlibrary loans of journal articles. Many libraries find that it is less expensive to borrow copies of journal articles from rarely used journals than to subscribe to these journals. Four recent studies[8,9,10,11] have provided excellent library use and cost data that demonstrate the trade-off within libraries concerning borrowing versus purchasing decisions. Basically the results suggest that it is less expensive to borrow copies than to purchase an entire journal if there are fewer than about five to ten uses of the journal. The exact break-even points depend on such factors as the price of the journal and number of issues per year. An important factor is that several studies[12,13,14] suggest that a fairly large proportion of journals in libraries have fewer than five or ten uses for any given volume (about 30 percent and 45 percent, respectively). Thus if libraries had knowledge of use and made their decisions on an economic basis only, they would cancel a number of journal subscriptions.

LIBRARY COST ASSOCIATED WITH SCIENTIFIC AND TECHNICAL JOURNALS

There are four principal library activities related to scientific and technical journals: acquisition of new journals, annual maintenance of journals, storage of journal copies, and use of journals in libraries.

New Acquisitions by Libraries

Input: Library decision to subscribe to a journal.
Process: Placement of order, follow-up on order, cataloging of new title.
Output: A newly acquired title.

The acquisitions process in a library may be lengthy and time-consuming; it involves reviewing titles being considered for acquisition, determining publisher and price, preparing orders, invoices, and payments, cataloging the new title, setting up records for automated control systems, and other associated record keeping. These procedures are performed for each new title.

The number of new titles acquired in a given library is generally about 2 percent of the titles held. This is not the same as the increase in the number of subscription copies received since titles are also discontinued by libraries.

The costs associated with new acquisitions, as well as other library operating costs, have been taken from a study of the costs of owning, borrowing, and disposing of periodical publications conducted recently for CONTU.[15] This study developed costs for three libraries: the main library of a large university, a smaller academic main library, and a large special library. These have been combined to arrive at a 1977 average to be used in establishing costs. Like other library processing costs, this figure consists almost entirely of labor costs.

Annual Maintenance by Libraries

Input: Regular library subscription.
Process: Receipt of journal issues and preparation for access and use.
Output: Library subscription copies available for use.

A number of operations are performed on a recurring basis for titles currently held by a library. These include review of the subscription, renewal of the subscription, back issue orders, check-in of each issue, claiming, catalog update, shelving of new issues, binding, and serial list updating. The extent of these processes varies among libraries and according to the particular journal title involved, but on the average annual processing of a serial title costs an estimated $31 for 1977 and can be projected for the other years under consideration.[16] The number of titles processed is the number of library subscriptions held, with adjustment for multiple subscriptions.

Library Storage

Input: Subscription copies received by library.
Process: Retention of journals for possible later uses.
Output: Stored journal volumes.

Library storage covers the physical storage of individual journal issues and bound journal volumes. In order to identify all such costs incurred in a given year, the number of volumes acquired over the previous fifteen years is estimated. The results of one such calculation indicate that the number of volumes stored in 1975 was 11.70 times the number of subscriptions held in 1975. The estimated cost of library storage in 1977 is taken to be $0.30 per annual volume, where an annual volume is defined to include all issues of a title for a calendar year. These costs primarily reflect the physical plant (rental value or other estimate of occupancy cost), including utilities and maintenance. Estimates for other years are based on a projection of this figure.

Library Use

Input: Stored journal ready for use.
Process: Use of the journal in the library, circulation, interlibrary loan, or use of the National Periodicals System.
Output: Article in the hands of the user.

Library use is categorized into use of journals in the library, circulation of journals for use outside the library, and interlibrary loan (transfer of a journal or a photocopy of an article from a library that subscribes to it to another library where a patron has requested it). For future costs, uses of articles obtained from a National Periodicals System are also included in library use.

In-library use and circulation do not involve substantial library operating costs, since in many cases the library patron identifies the required journal article and physically locates it without any involvement of the library staff. Some uses may involve staff labor in terms of reference help, reshelving of materials, or check-out of materials. Interlibrary loan is a more costly operation, since it requires recording a user's request, identifying a potential lending library, transmitting a request, making additional requests if necessary, receipt of the item, notifying the patron, and returning the borrowed journal. These operations are performed by the borrowing library, and still others are

Table 5.7   Unit costs of library activities associated with journals, 1977.

Activities	Unit Cost
Processing	
Acquisitions	$95.91 per new journal[a]
Annual maintenance	$30.92 per journal[a]
Annual storage	$6.00 per journal
Weeding	$0.90 per journal
Use	
Internal use	
and circulation	$2.00 per use
Interlibrary loan	$11.60 (borrowing library) +
	$8.40 (lending library) = $20.00 per loan

Source: Projected from G. Williams, *Library Cost Models: Owning versus Borrowing Serial Publications* (Rockville, Maryland: Westat Research, 1968).
[a]Exclusive of subscription price.

required of the lending library. In all, the estimated total cost of a completed interlibrary loan in 1977 is about $20.[17] By contrast, the cost of one in-library use is estimated at about $2, and the cost of a journal circulation at slightly more than that.

Costs associated with the National Periodicals System include the costs of acquisition and storage of articles, request costs, and retrieval and distribution costs. Acquisition and storage costs depend on the number of articles acquired and the technologies used. In the early years of the development of the system, we hypothesize, there will be both paper and electronic storage. Transmission of requests and article copies may be either by mail or electronically, with different costs associated with each.

There are few available data on the relative mix of in-library circulation, and interlibrary loan uses in libraries. Circulation is relatively infrequent, since many libraries do not circulate journals. Interlibrary loan of scientific and technical journals has recently been estimated at about 4 million loans in 1977. Total journal use from libraries is estimated in this study as 36 million uses for the same year. These estimates are combined into the costs for in-library, circulation, and interlibrary loan use of journals, using the unit costs given in Table 5.7. The total costs are found by applying the unit acquisition cost to the estimated number of new journals, maintenance, storage, and weeding unit costs to the estimated number of library subscriptions, and the respective unit costs to the estimated number of uses from internal copies and interlibrary loans. These costs are summarized in Table 5.8.

Table 5.8 Estimated total cost of library activities associated with journals, 1975–1985 (in millions).

Source of Cost	1975	1977	1980	1985
Acquisitions	$14	$15	$19	$25
Maintenance, storage, weeding	250	274	337	447
Use	115	114	205	349
Total	379	433	561	821

FEDERAL CONTRIBUTION TO LIBRARIES

The federal contribution to library expenditures comes primarily through two channels: expenditures of federal libraries and federal grants to other types of libraries. Federal scientific and technical library expenditures are estimated at $193 million for 1977, which includes costs associated with all types of materials and of federal subscriptions to journals. The $193 million figure is thus reduced to $150 million for nonmaterial expenditures and further to $22 million for processing of U.S. scientific and technical journals. This last figure is taken as the relevant federal contribution.

Direct subsidies to libraries come from such programs as those provided by the Library Services and Construction Act and the Higher Education Act. Few of these projects specifically involve journals, with the exception of some of the networking activities, so direct subsidies are not included in our accounting of the federal contribution.

The federal government also indirectly supports libraries, and thus journal use, through the funding of research overheads. The magnitude of federal research funding is $23.5 billion, and it can be estimated that about 7 percent, or $235 million, goes to libraries through the overhead rates. Again assigning a proportion of this to U.S. scientific and technical journals, a figure of $34 million is arrived at.

SECONDARY SERVICES ACTIVITIES

Abstracting and Indexing Services

Abstracting and indexing services (A&I) play an important role in scientific and technical information transfer by providing the intellectual analysis and organization of the scientific literature that gives scientists and engineers the effective access they need for their research activities. Although they concentrate primarily on the journal literature,

A&I services treat all literature to some extent and provide scientists with current awareness of materials relevant to their fields of interest, as well as guides for retrospective search of the world's literature.

A&I services are provided by organizations widely varying in the nature of their sponsors and in their sources of funds but generally producing similar products. The traditional product is a periodic publication listing items considered, often with abstracts, and accompanied by appropriate indexes. Abstracts may be indicative, informative, evaluative, or any combination of these types. Indexes also vary in type as well as comprehensiveness; the principal forms used are descriptive cataloging, alphabetic or hierarchic subject, keyword, and citation indexes. In the development of the traditional product, a number of organizations have created machine-readable data bases containing similar information that can be used to provide index services directly to persons searching on-line. The growth of such services has been substantial in recent years, which may suggest correspondingly less emphasis on the printed A&I product in the years ahead as printed materials are replaced by on-line searches.

A number of attempts have been made over the years to identify various categories of organizations providing A&I services, going back at least to the first edition of *Index Bibliographicus* published in 1925. Of more recent interest, *A Guide to U.S. Indexing and Abstracting Services in Science and Technology,*[18] was published in 1960 by the National Federation of Abstracting and Indexing Services (NFAIS).* This compilation contained about five hundred entries† and was limited to U.S. services. Three hundred and sixty-five of these services were repeated in the 1963 *World Guide to Abstracting and Indexing Services in Science and Technology,*[19] another NFAIS publication. This document contained 1,855 titles from forty countries, chosen from 3,115 titles examined. In 1965 and again in 1969, the Fédération internationale de documentation (FID) published guides to abstracting (but not indexing) services in science, technology, medicine, and agriculture and also in social sciences and humanities. The 1969 edition contained 816 entries in the science and technology volume and 180 entries in the social sciences volume. A comparison of it with that of the 1963 NFAIS *World Guide* indicates an increase of at least 30 percent in the number of science and technology abstracting and indexing services over the six-year period, and a combination of the two lists identifies 2,671 unique entries representing worldwide A&I services in all subject fields.

*NFAIS in origin was a federation of science A&I services. The great majority of its members' coverage is still science and technology.
†Distinct services by the same sponsoring organization were counted separately.

Table 5.9 Number of abstracting and indexing services,
1960–1975.

Type of Organization	1960	1963	1969	1975
U.S. science and technology	500	365		330
World science and technology		1,855	816[a]	670
All fields			996	2,100

Sources: National Federation of Abstracting and Indexing Services, *A Guide to U.S. Indexing and Abstracting Services in Science and Technology* (Washington, D.C.: National Federation of Abstracting and Indexing Service, 1960); National Federation of Abstracting and Indexing Services, *A Guide to the World's Abstracting and Indexing Services in Science and Technology* (Washington, D.C.: National Federation of Abstracting and Indexing Services, 1963); Fédération internationale de documentation, *Abstracting Services. I. Science, Technology, Medicine, Agriculture, II. Social Services and Humanities* (The Hague: Fédération Internationale de Documentation, 1969).
[a]Indexing services not counted.

In an effort to merge and update data contained in these previous directories and to create a base for ongoing data collection, FID and NFAIS carried out a joint project to prepare a machine-readable inventory of world abstracting and indexing services. The data base was completed in 1976. Included in the record for each service are ninety-one elements describing the service. Services listed were selected primarily from existing directories, according to criteria established by an advisory committee. Preliminary results of this project indicate a final data base containing descriptions of approximately twenty-one hundred basic A&I services plus approximately two hundred machine-readable services, some of which are cross-referenced to corresponding basic services. By conservative estimates, about a thousand of the services are scientific and technical in nature. One-third of these are based in the United States.

Combining the known information, Table 5.9 suggests some tentative conclusions. Indications are that the United States accounts for a relatively small percentage (20 percent in 1963) of the world's A&I services in science and technology. About half of the services deal with the scientific and technical literature. With respect to the total number of services, the 1975 estimate of 2,100 world A&I services seems the most reliable. Substantially less growth is seen in the 1969–1975 period for scientific and technical services than the 30 percent estimated for 1963–1969.

Also of interest is the number of primary publications processed by the A&I services. These data will be included in the new World Guide and are partially available through the annual statistics of NFAIS. Data

Table 5.10   Items processed by U.S. abstracting and indexing
services, 1960–1985 (in thousands).

Year	Items Processed by U.S. NFAIS Members	S&T Items Processed by U.S. A&I Services
1960	588	1,058
1961	676	1,217
1962	713	1,283
1963	756	1,361
1964	795	1,431
1965	884	1,591
1966	986	1,775
1967	996	1,793
1968	1,135	2,041
1969	1,226	2,207
1970	1,257	2,263
1971	1,376	2,477
1972	1,489	2,680
1973	1,445	2,601
1974	1,478	2,660
1975	1,623	2,921
Projections		
1976	1,681	3,025
1977	1,751	3,152
1978	1,821	3,278
1979	1,891	3,404
1980	1,960	3,528
1981	2,029	2,653
1982	2,099	3,779
1983	2,169	3,904
1984	2,339	4,030
1985	2,308	4,155
Percent change		
1960–1965	50	50
1965–1970	42	42
1970–1975	29	29
1975–1980	21	21
1980–1985	18	18

Sources: National Federation of Abstracting and Indexing Services, *Member Service
Statistics* (Philadelphia, Pennsylvania: National Federation of Abstracting and
Indexing Services, 1974, 1975, and 1976); King Research, Inc.
Note: Each journal article, report, or thesis is counted as an item, and generally
each separately authored chapter of a book is so counted.

for the NFAIS member services, presented in Table 5.10, show steady
growth over the years. Projections beyond 1977 indicate increases of
about 4 percent annually to 1985.

It has been estimated that the number of items processed by NFAIS members represent about half of the total U.S. volume. A major organization that is not a member of NFAIS is the Institute for Scientific Information (ISI); it processed 113,000 items in 1961 and 419,000 in 1974.

Bibliographic Data Bases

Bibliographic data bases, initially created as by-products of the A&I publication process, are increasingly becoming an information product in their own right. Covering in general the same literature as do the traditional products, they allow in-depth searching of large volumes of information by computer. As the development of bibliographic data bases has increased, new and related service organizations have been established to facilitate use of them. These new service organizations include vendors or suppliers, who process the data bases and provide computer access to them, and brokers, who perform computer searches for users.

Information about data bases has been compiled increasingly as the industry develops, and today there are several sources of continuing information. In recent years there have also been several studies pertaining to particular aspects of data-base creation and use. Among the sources of continuing information are *Computer-Readable Bibliographic Data Bases—A Directory and Data Sourcebook,*[20] Kruzas's *Encyclopedia of Information Systems and Services,*[21] the *Annual Review of Information Science and Technology,*[22] the *Bulletin of the American Society for Information Science* and *NEWSIDIC,* the newsletter of the European Association of Scientific Information and Dissemination Centers. Special studies in the area have recently been done by the System Development Corporation[23] and by Applied Communications Research.[24]

The most comprehensive source of information on data bases is the publication *Computer-Readable Bibliographic Data Bases—A Directory and Data Sourcebook.*[25] This directory, published in 1976, was compiled by Williams and Rouse from a data base on data bases, maintained at the Information Retrieval Research Laboratory (IRRL) at the University of Illinois. The file is continuously updated, and publication updates are expected every six months, with complete new indexes generated annually. Data bases are included in the directory if they contain bibliographic or bibliographic-related information, if they are commercially or publicly available through either the producer or a processing center, and if they are designed or used primarily for information retrieval purposes rather than in library processing. For

Table 5.11   Subject breakdown of bibliographic data bases produced in the United States and other countries, 1975.

Subject	No. of Data Bases		
	United States	Other Countries	Total
Science and technology	88	98	186
Medicine and life sciences	15	13	28
Science and technology related	41	3	44
Total S&T	144	114	258
Nonscience or technology	35	8	43
Total data bases	179	122	301

Source: Computed from M. E. Williams and S. H. Rouse, *Computer-Readable Bibliographic Data Bases—A Directory and Data Sourcebook* (Washington, D.C.: American Society for Information Science, 1976).

each data base, the directory lists from 5 to 370 elements of data, including such information as generator, producer, distributor, availability, price, size, frequency, scope, subject matter, and type of material covered.

The directory contains information about 301 data bases, 179 produced in the United States and 122 in other countries. A breakdown of the data bases by subject matter is presented in Table 5.11. About 80 percent of U.S. data bases wholly or in part cover scientific and technical publications, as compared to 93 percent of data bases in other countries. This reflects the relative maturity of the data-base industry in the United States since the established pattern has been first to develop scientific and technical data bases and later to go on to other subject fields. Of the 144 U.S. scientific and technical data bases, 61 percent specifically cover the sciences and technology (excluding medicine and the life sciences), 10 percent cover medicine and the life sciences, and the remaining 29 percent are multidisciplinary but include some scientific and technical material.

The U.S. scientific and technical data bases can also be broken down by the sector—government or private—producing each of them. This is done in Table 5.12, indicating both the number of data bases and the number of records in them. About one-third of the data bases are produced in the government sector and two-thirds in the private sector. Considering the total number of records in the data bases at the end of 1975, the government sector accounts for a smaller percentage of 22 percent, with 78 percent produced in the private sector. On the average, there were slightly more than 300,000 records in each U.S. scientific and technical data base in December 1975 and an average of

Table 5.12 Number, size, and growth of U.S. scientific and technical bibliographic data bases, 1975.

	Sector			
	Government	Private Not-for-Profit	Private For-Profit	Total
Number of data bases	49 (34%)	57 (40%)	38 (26%)	144 (100%)
Number of records, December 1975 (millions)[a]	8.5 (22%)	19.4 (51%)	10.1 (27%)	38.0 (100%)
Average number of records, December 1975	206,000	418,000	335,000	306,000
Number of records added annually (millions)[a]	1.0 (16%)	3.4 (53%)	2.0 (31%)	6.4 (100%)
Average number of records added	26,000	77,000	68,000	58,000

Source: Computed from M. E. Williams and S. H. Rouse, *Computer-Readable Bibliographic Data Bases—A Directory and Data Sourcebook* (Washington, D.C.: American Society for Information Science, 1976).
[a]Includes only data bases that provided information on the number of records (124) or the number of records added (111).

58,000 records added in 1976. These figures vary substantially by the individual data base and also by sector. Government sector data bases are, on the average, smaller than those produced in other sectors.

Overall more than 6 million records are added annually to the scientific and technical data bases. This represents coverage of a wide range of literature, both U.S. and foreign, but does include substantial overlap. It is impossible to determine the actual extent of coverage. It does appear, however, that a large proportion of this literature is being covered.

Data-Base Searching

Since bibliographic data bases in general cover the same literature as do the A&I products, they allow in-depth searching of large volumes of information by computer. As these data bases have developed, related service organizations have been established to facilitate their use by direct access to them and by intermediaries. The strong growth of bibliographic searching in the United States between 1965 and 1975 is shown in Table 5.13, reaching nearly 1.5 million on-line searches in 1977, rapid growth that is expected to continue.

Additional data on bibliographic data-base searching were obtained from a survey of scientists and engineers regarding their use of journals and also of bibliographic and numeric data bases. About 6 percent reported having performed or used the output of a bibliographic data base within the last two weeks. This percentage varied considerably by field of science (Table 5.14). There were relatively few uses by Computer Scientists and Engineers, while 20 and 34 percent of Life Scientists and Environmental Scientists, respectively, had performed at least one search. If each scientist reporting some search activity is assumed to have made only one search, Table 5.14 suggests an estimated 149,000 searches in a two-week period, or over 3.8 million searches a year. These figures indicate that data-base searching activity may be even greater than was projected.

Respondents were also asked who performed the actual computer search and who examined the output. Eighty percent of the time, the search was conducted by an intermediary, with the scientist evaluating the output. Fifteen percent of all searches involved only the scientist, both as searcher and evaluator. In the remaining 5 percent, the survey respondent performed the search but did not review the output. The average time spent by survey respondents in these activities was 100 minutes over the two-week period considered, or about 2 percent of the scientists' time. This compares with about 190 minutes spent

Table 5.13   Estimated number of on-line data-base searches, 1968–1985.

Year	Searches (000)
1968	10
1969	50
1970	120
1971	200
1972	300
1973	450
1974	610
1975	1,000
Projections	
1976	1,220
1977	1,440
1978	1,900
1979	2,210
1980	2,500
1981	2,750
1982	2,980
1983	3,230
1984	3,470
1985	3,700
Percent change	
1970–1975	733
1975–1980	150
1980–1985	48

Sources: M. E. Williams, "United States Versus European Use of Data Bases," *NEWSIDIC* (Information Bulletin of EUSIDIC) **14**:11 (1974); King Research, Inc.

Table 5.14   Bibliographic data-base use, 1977.

Field of Science	Number of Scientists Searching (000)	Proportion of All Scientists (%)
Physical Sciences	16.0	7.7
Mathematics	4.7	4.8
Computer Sciences	0.6	0.4
Environmental Sciences	25.5	33.8
Engineering	2.2	0.2
Life Sciences	56.2	20.3
Psychology	11.5	10.7
Social Sciences	17.9	7.9
Other Sciences	14.6	14.6
All Fields	149.2	6.4

Source: King Research, Inc. User Survey.

Table 5.15   Numeric data-base use, 1977.

Field of Science	Number of Scientists Searching (000)	Proportion of All Scientists (%)
Physical Sciences	64.4	31
Mathematics	20.4	21
Computer Sciences	79.1	49
Environmental Sciences	15.9	21
Engineering	270.6	25
Life Sciences	58.1	21
Psychology	27.9	26
Social Sciences	135.7	60
Other Sciences	13.0	13
All Fields	685.1	29

Source: King Research, Inc. User Survey.

reading in a two-week period. In the case of the relatively small group of scientists using bibliographic data bases, then, the investment of time is considerable.

Survey respondents were also asked about their use of numeric data bases. Here 29 percent indicated some use of such data bases within the most recent two weeks. By field, this percentage rose as high as 49 percent in the Computer Sciences and 60 percent in the Social Sciences (Table 5.15). Respondents indicated heavy use of environmental and engineering data bases (49 and 56 percent, respectively, of those who used data bases at all), and lesser use of social characteristics data bases (27 percent of those who used data bases). Social Scientists used both social characteristics and environmental (including geographic) data bases heavily, as did Environmental Scientists. The heaviest use of engineering data bases was, not surprisingly, by Engineers. Average time spent using data bases was over ten hours within the last two weeks.

Cost of Secondary Services

A&I activity is described as follows:

Input: Journal article information* (available through subscription, publishers' reprints, references from other scientists, and so forth).

*The focus of this book is on journal articles although secondary services cover other scientific and technical publications.

Process: Abstracting, indexing, creating, and distributing indexes.
Output: Bibliographic access guides to the journal literature.

This category is concerned with the development, production, and distribution of A&I service products, including printed indexes and machine-readable bibliographic data bases. For 1977, it has been estimated that about 3.2 million scientific and technical periodical items and 1.2 million scientific and technical journal items (37 percent) were processed by U.S. A&I services. At an estimated processing cost of $36.50 per item, the total 1977 A&I cost for these journals is $45.3 million. This total can be viewed as $0.19 per article use or $120 per article published, keeping in mind that not all article uses involve the use of A&I services. For lack of an alternative assumption, the $120 per article figure is used in our calculations for 1977. Similar estimates are derived for 1980 and 1985.

A&I activity would clearly be affected by electronic publishing processes, possibly by increased coverage, more automation of processes, and decreased production of paper indexes. The overall magnitude of such changes has not been estimated for this study. Decreased overlap would also influence cost but might ill serve specialized audiences.

Data-Base Searching Costs

To consider the use to which the available data bases are put, the System Development Corporation conducted surveys of both managers and searchers in over five hundred organizations using the services of ten major suppliers of on-line bibliographic services.[26] Among the topics covered were the level of on-line use, training, selection, and use of on-line systems and data bases, costs, and effects. Overall it was estimated that more than 500,000 on-line searches were conducted within the study domain during 1975 and that at least 120,000 hours were spent in conducting these searches.

Other findings of the study identified managers' estimates of the cost of an on-line search as ranging from $1 to $99, with a mean of $24 and a median of $17. This cost generally included charges for computer, communications, and off-line printing, and in about half of the cases it included the cost of staff time and terminal rental. Computer and printing charges vary substantially depending on the data base and system used. Staff time includes both actual time at the terminal (mean nineteen minutes, median fifteen minutes) and also related preterminal and postterminal work by search staff (means twenty-one and eighteen

Table 5.16 Estimated costs of secondary services associated with
scientific and technical journals, 1975–1985
(in millions).

Type of Service	1975	1977	1980	1985
Abstracting and indexing[a]	$42	$45	$51	$57
Data-base searching	71	107	199	334
Total	113	152	250	391

[a]Including paper and electronic publishing.

minutes, respectively, medians fifteen and ten minutes). In general, a
search seems to have been interpreted as the activities required to fulfill
one request, possibly including searches of multiple data bases and/or
multiple searches of one data base.

A similar definition of *search* was used in an Applied Communications
Research project[27] dealing with the cost of on-line bibliographic
searching. This project was part of a larger evaluation concerned with
the use of on-line search systems in public libraries and concerned itself
with the costs of searching in four public libraries in the San Francisco
Bay area. These libraries were using the Lockheed DIALOG retrieval
service on an experimental basis during the study period. Costs
identified by ACR totaled $29 per search, including $17 for data-base
charges, $6 for off-line printing, and $5 for staff time. Overhead and line
charges were not included. Staff time was equally divided between work
at the terminal and other activities.

These studies suggest average costs to the user of about $35 per
on-line search in 1975. Although the actual cost may be somewhat
larger due to partial subsidization of the search services, it is
nonetheless clear that the cost of on-line searching has decreased
significantly since its early days. This trend can be expected to continue
as the volume of on-line searching increases.

The total costs of secondary services are summarized in Table 5.16.
The total costs have been increasing dramatically, which reflects the
rapid growth in these activities.

FEDERAL CONTRIBUTION TO
SECONDARY SERVICES

A survey of federal A&I services, many producing data bases as well
as printed publications, was conducted to determine the magnitude of
federal participation in abstracting and indexing and the associated

costs. The universe for the survey was the fifty-two publications listed in the *1977 Serials Supplement to the Monthly Catalog of U.S. Government Publications.* After initial inquiries, this number was reduced to thirty-one active services. A number of publications, such as *Neoplasma Abstracts, Aerospace Medicine and Biology,* and *Aeronautical Engineering,* which are subsets of larger files, were counted as a part of the parent file. Data on fifteen additional services were obtained from a previous survey. Together the forty-six services surveyed processed an estimated total of 1.5 million items, with data for the individual services ranging from 90 for *Computer Program Abstracts* to 160,000 for *Index Medicus.*[28] Expenditures were not given in all cases, but those supplied averaged about $30 per item. Applying this figure to the 1.5 million items processed yields a total federal contribution of $45.0 million.

For the most part, the federal A&I services surveyed were completely funded through agency appropriations. In a few instances, however, appropriations were supplemented with additional funds through the sale of annual subscriptions, searches, tapes, data bases, and other services.

It is estimated that 37 percent of data-base search costs (approximately $40 million) are incurred by federal employees or federally funded researchers. This makes the total federal contribution to U.S. scientific and technical journal expenditures $56 million for libraries and $85 million for secondary services, for a total of $141 million.

REFERENCES

1. A. T. Kruzas, ed., *Encyclopedia of Information Systems and Services,* 2d ed. (Ann Arbor, Michigan: Anthony T. Kruzas Associates, 1974).
2. *American Library Directory* (New York: Bowker, 1962–1976).
3. Association of Research Libraries, *American Library Statistics* (Washington, D.C.: Association of Research Libraries, 1960–1972).
4. Association of Research Libraries, *ARL Statistics* (Washington, D.C.: Association of Research Libraries, 1974–1976).
5. M. A. Drake, *Academic Research Libraries: A Study of Growth* (West Lafayette, Indiana: Purdue University, 1977).
6. *Bowker Annual of Library and Book Trade Information,* 4th–21st eds. (New York: Bowker, 1961–1976).
7. Department of Health, Education and Welfare, National Center for Educational Statistics, *Survey of Federal Libraries* (Washington, D.C.: Government Printing Office, 1972).

8. V. E. Palmour, M. C. Bellassai, and L. M. Gray, *Access to Periodical Resources: A National Plan* (Washington, D.C.: Association of Research Libraries, 1974).
9. V. E. Palmour, E. C. Bryant, N. Caldwell, and N. W. Gray, *A Study of the Characteristics, Costs and Magnitude of Interlibrary Loan in Academic Libraries* (Westport, Connecticut: Greenwood Publishing Company, 1972).
10. V. E. Palmour, M. C. Bellassai, and R. R. V. Wiederkehr, *Costs of Owning, Borrowing, and Disposing of Periodical Publications* (Arlington, Virginia: Public Research Institute, 1977).
11. Task Force on a National Periodicals System, *Plan for a National Periodical Program* (Washington, D.C.: National Commission on Libraries and Information Science, 1977).
12. C.-C. Chen, "The Use Pattern of Physics Journals in a Large Academic Research Library," *Journal of the American Society for Information Sciences* 23(4):254–265 (1972).
13. A. Kent, K. L. Montgomery, J. Cohen, J. G. Williams, S. Bulick, R. Flynn, W. N. Sabor, and J. R. Kern, *"A Cost-Benefit Model of Some Critical Library Operations in Terms of Use of Materials"* (Pittsburgh, Pennsylvania: University of Pittsburgh, 1978).
14. G. Williams, *Library Cost Models: Owning Versus Borrowing Serial Publications* (Rockville, Maryland: Westat Research, 1968).
15. Palmour, Bellassai, and Wiederkehr, *Costs of Owning, Borrowing, and Disposing.*
16. Ibid.
17. Ibid.
18. National Federation of Abstracting and Indexing Services, *A Guide to U.S. Indexing and Abstracting Services in Science and Technology* (Washington, D.C.: National Federation of Abstracting and Indexing Services, 1960).
19. National Federation of Abstracting and Indexing Services, *A Guide to the World's Abstracting and Indexing Services in Science and Technology* (Washington, D.C.: National Federation of Abstracting and Indexing Services, 1963).
20. M. E. Williams and S. H. Rouse, *Computer-Readable Bibliographic Data Bases—A Directory and Data Sourcebook* (Washington, D.C.: American Society for Information Science, 1976).
21. Kruzas, *Encyclopedia.*
22. C. A. Cuadra, ed., *Annual Review of Information Science and Technology,* vols. 1–12 (Washington, D.C.: American Society for Information Science, 1966–1977).
23. J. Wanger, C. A. Cuadra, and M. Fishburn, *Impact of On-Line Retrieval Services: A Survey of Users, 1974–75* (Santa Monica,

California: System Development Corporation, 1976).
24. O. Firschein, R. K. Summit, and C. K. Mick, "Use of On-Line Bibliographic Search in Public Libraries: A Retrospective Evaluation," *On-Line Review* 2(1):41–55 (1978).
25. Williams and Rouse, *Computer-Readable Bibliographic Data Bases.*
26. Wanger, Caudra, and Fishburn, *Impact.*
27. M. D. Cooper and N. A. DeWath, "The Cost of On-Line Bibliographic Searching," *Journal of Library Automation* 9(3):195–209 (1976).
28. B. L. Wood and C. G. Schueller, *Results of the Federal Scientific and Technical Information Survey* (Rockville, Maryland: King Research, 1977).

Chapter 6

Use of Scientific and Technical Journals

The final state of the communication cycle involves the direct use of a journal article by the individual scientist or engineer. In a sense, all other activities in the communication cycle have led up to this event, and from it, if all goes well, the cycle will be repeated at a higher level. Thus use can be seen as the most critical activity of the communication cycle. Unfortunately, it is one of the most difficult to monitor and quantify. This chapter presents findings, primarily from our recent survey of scientists and engineers, about current levels of use of the journal literature.

In our research, we make the implicit assumption that reading of a scientific and technical journal article implies its use, or at least that use does not occur without reading. We define reading as "going beyond the contents, title and abstract, to the body of the article." A major exception to the assumptions made is use of a journal article through discussion with a colleague who has read it (without the second user actually reading it), but this secondary level of use is not included in our data. Similarly we do not attempt to identify quantitatively multiple uses of a given article resulting from a single reading. Including these and similar categories of use in our statistics would clearly result in a level of use higher than the level of reading shown in this chapter.

DESCRIPTION OF SCIENTIFIC AND TECHNICAL JOURNAL USE

A substantial number of studies have provided estimates of the amount of time spent reading the literature generally and scientific and technical journals specifically, but very few have addressed the question of readership or the purposes of reading journal articles. This leaves a large gap in the general description of scientific and technical communication. Specific models of the cost and flow of journal information are greatly dependent on readership and use data.

Furthermore the purposes of reading journal articles present a useful surrogate measure of benefit that can be compared to journal system cost.

In the 1960s an excellent study of information flow (primarily in psychology) was performed by the American Psychological Association under an NSF grant.[1] Also about that time Case Institute conducted a readership study for NSF in the area of physical sciences.[2] The only very recent studies of readership of which we are aware are one done by Machlup et al.[3] in the area of economics and one by King Research (KRI).[4] in the area of cancer research. However, all lack data for our purpose of modeling and analysis. Thus one of the principal efforts during 1978 was to obtain journal readership and use data with a national survey of scientists and engineers.

Readership of Scientific and Technical Journals

We took a random sample of scientists from a large number of professional society membership lists and from a list of engineers and computer scientists provided by R. L. Polk and Co. This sample was supplemented by a random sample of 1977 journal article authors. Those sampled from these three sources were sent questionnaires, and there was a telephone follow-up. This process yielded a total usable sample of 2,350 responses (or a response rate of 89 percent for authors and 61 percent for nonauthors).* From these observations, data about extent of authorship, journal readership, article use, subscription behavior, and use of bibliographic and numeric data bases were gathered and analyzed.†

The scientists whom we surveyed were asked first whether they were currently engaged in research or development activities. If not, they were not included in the remaining results. If they were, they were asked primarily about journal articles that they had read in the last month. Scientists who did not read in the last month were assumed not to be heavy readers of the journal literature and were classified as nonreaders; this group made up about 12 percent of the population overall (Table 6.1). As might be expected, the proportion of readers is significantly higher than average in the fields of Environmental Sciences, the Life Sciences, and Psychology and lower than average in

---

*For calculating response rate, a nonauthor is a person whose name was drawn from one of the lists rather than from a journal. Response rates given include both mail survey (long form) and telephone follow-up responses (short form).
†The estimates for the observations were weighted to adjust for multiple society membership and disproportionate sampling from various strata.

Table 6.1   Readers and nonreaders of scientific literature, by field of science, 1977 (in thousands).

Field of Science	Readers (No.)	(%)	Nonreaders (No.)	(%)	Total (No.)	(%)
Physical Sciences	185.5	89	22.3	11	207.8	100
Mathematics	74.5	77	22.8	23	97.3	100
Computer Sciences	138.5	86	22.9	14	161.4	100
Environmental Sciences	75.1	99	0.4	1	75.5	100
Engineering	947.6	88	134.7	12	1,082.3	100
Life Sciences	266.5	96	10.2	4	276.7	100
Psychology	102.4	95	4.9	5	107.3	100
Social Sciences	207.8	92	18.4	8	226.2	100
Other Sciences	55.9	56	44.0	44	99.9	100
All Fields	2,053.7	88	280.7	12	2,334.4	100

Source: King Research, Inc. User Survey.

Mathematics and Other Sciences. Another number to be noted in Table 6.1 is the total number of scientists and engineers, 2.3 million. This number reflects an adjustment, based on our survey results, from the total estimates of U.S. scientists and engineers (2.8 million) down to those who are actually involved in research and development, either directly or through management, supervision, teaching, or training. Estimates of the total scientific community, by field of science, were shown in Table 3.1.

The estimated total amount of readership by R&D-related scientists and engineers is given in Table 6.2. The total number of articles read is 244 million, or an average of 105 articles per scientist per year. Estimates by field are in line with those determined for individual fields in the studies mentioned above. Overall there is no previous basis for comparison except for one earlier estimate of 352 million made by KRI for 1975.[5] This was based on incomplete data and some very broad assumptions, whereas the current estimate resulted from a refined methodology applied across all the fields of science. Thus we view the current estimate as the first one of total readership based on a sample survey. However, we caution that only 61 percent of nonauthors responded to the survey and nonresponders may read less than responders even though our estimates assume equal readership by both groups.

Among the fields of science,* the data in Table 6.1 show about one-third of all readings of scientific and technical articles being made by Engineers, and 10 to 20 percent each by Life Scientists, Physical

*Estimates throughout this chapter for individual fields of sciences are sometimes based on as few as fifty observations.

Table 6.2   Estimated total and average number of articles read in
one year, by field of science, 1977.

Field of Science	Articles Read in One Year (millions)	No. of Scientists or Engineers (000)	Average Number of Articles Read in One Year
Physical Sciences	39.3	207.8	190
Mathematics	5.9	97.3	60
Computer Sciences	10.9	161.4	70
Environmental Sciences	8.2	75.5	110
Engineering	82.6	1,082.3	80
Life Sciences	49.7	276.7	180
Psychology	11.5	107.3	110
Social Sciences	28.3	226.2	120
Other Sciences	7.6	99.9	80
All Fields	244	2,334	105

Source: King Research, Inc. User Survey.

Scientists, and Social Scientists. Later these numbers will be compared
with the numbers of article copies distributed. From the table's
indication of average number of articles read per scientist, we can see
that the high number of readings by Engineers occurs despite the low
average number of readings (80 per scientist per year) and only because
of the large size of the engineering population. Both total number of
scientists and average readers are high for the Physical and Life
Sciences (190 and 180 readings, respectively). The lowest average levels
of reading are found in Mathematics (60 readings per scientist per year)
and Computer Sciences (70 readings per scientist per year). See
Table 6.2.

The user survey showed that of the articles read at all, 39 percent*
were read with great care, 57 percent were read with attention to the
main points, and 4 percent were read just to get the idea. (Table 6.3).
The thoroughness of reading seems to vary somewhat among fields,
with Engineers reading more articles with great care and
Mathematicians, Psychologists, and Social Scientists reading a higher
percentage of articles just to get the idea. A large proportion of readings
by Physical Scientists and Environmental Scientists are of intermediate
depth. One survey of chemists[6] revealed that about 48 percent of those
surveyed read up to 25 percent of an article, 20 percent read up to
50 percent of an article, and 12 percent read up to 75 percent of an
article. This does not seem much out of line with our user survey results

*The respondents were told that reading meant going beyond the table of contents, title,
and abstract to the body of the article.

Table 6.3 Proportion of reading by depth, by field of science, 1977 (in percent).

Field of Science	Articles Read with		
	Great Care	Attention to Main Points	Just to Get the Idea
Physical Sciences	16	83	1
Mathematics	27	57	16
Computer Sciences	33	66	1
Environmental Sciences	10	85	5
Engineering	49	50	1
Life Sciences	40	55	5
Psychology	28	57	15
Social Sciences	39	43	18
Other Sciences	23	77	a
All Fields	39	57	4

[a]Less than .005.

involving Physical Scientists. The economic journal readership survey performed by Machlup and Leeson[7] also showed similar results to those found in Social Sciences, where economists are classified.

Methods of Access to the Journal Literature

To obtain a journal article for reading, scientists and engineers must go through a two-stage process. First is the intellectual identification of a particular article as appropriate to their needs; this may occur with or without the article being physically present. Common methods of identification are through browsing, through a colleague, and from a citation in a primary or secondary publication. If this does not involve a physical copy of the article, physical access becomes the second step of the process. Here two variables are relevant: the location of the article (library, colleague's collection, author's file, and so on) and its form (journal issue, photocopy, microform, reprint or preprint, or other).

Procedures used by scientists for identifying articles are indicated in Table 6.4, and sources of access are shown in Table 6.5. Overall the most predominant means of identifying an article is browsing, occurring with 40 percent of the article readings, but identification through a printed index is also quite frequent. The latter data, which suggest that nearly 60 million articles that were read in 1977 were identified through printed indexes, provide an indication of the level of use of these publications. Results for identification through computer search indicate

Table 6.4   Procedures for finding articles read, by field of science, 1977 (in percent).

Field of Science	Browsing in an Issue	From Another Person	Cited in an Article	Cited in Printed Index	Computer Search	Other
Physical Sciences	49	3	12	35		2
Mathematics	24	38	21	11		6
Computer Sciences	34	33	a	33	a	a
Environmental Sciences	47	29	1	a		23
Engineering	34	17	16	33	a	a
Life Sciences	46	18	5	10	2	18
Psychology	60	14	11	9	a	6
Social Sciences	42	23	5	15		15
Other Sciences	78	5	3	1		14
All Fields	40	18	11	24	0.2	6

[a]Less than 0.005.

Table 6.5   Source of articles read, by field of science, 1977 (in percent).

Field of Science	Own Subscription Copy	Library Copy	Author	Coworker	Other
Physical Sciences	76	22	1	1	a
Mathematics	30	50	8	10	3
Computer Sciences	34	33	33	a	a
Environmental Sciences	65	15	19	1	a
Engineering	83	1	16	a	a
Life Sciences	50	26	7	15	1
Psychology	62	25	a	7	6
Social Sciences	57	24	5	8	6
Other Sciences	87	8	2	2	1
All Fields	69	14	12	4	1

[a]Less than 0.005.

a much lower level of a half-million readings for which a computer data base was the identification source.

Table 6.5 shows a great dependence on individual subscription copies; these are the source for 69 percent of readings. It seems reasonable that these would include most of the articles identified by browsing, as well as some identified by other means. Fourteen percent of articles read are library subscription copies and reprints or preprints from the author. The proportion of articles obtained from colleagues

Table 6.6 Forms of articles read, by field of science, 1977
(in percent)

Field of Science	Actual Issue	Reprint	Preprint	Photocopy	Other
Physical Sciences	96	1	a	3	a
Mathematics	61	1	7	31	a
Computer Sciences	67	a	32	a	a
Environmental Sciences	79	19	a	1	a
Engineering	83	a	a	17	
Life Sciences	82	8	a	10	
Psychology	90	a		10	
Social Sciences	81	6	a	10	3
Other Sciences	95	3		2	
All Fields	83	2	3	12	a

[a]Less than 0.005.

(4 percent overall) seems surprisingly low in view of the 18 percent of identifications made through colleagues.

There are some interesting variations among the fields of science in methods of finding articles and in sources used. The fields differ in both the number of methods and sources used extensively and in the distributions among these. In Psychology and the Other Sciences, for example, browsing is the only major identification source, while in Mathematics, Computer Sciences, and Engineering there are at least three major identification sources. Dependence on citations is particularly high among Mathematicians, while Physical Scientists, Computer Scientists, and Engineers used printed indexes for more than one-third of their article identifications. In keeping with these findings, there is also more variability of article sources for Mathematicians, Computer Scientists, and Life Scientists. Dependence on subscription copies as an access source is high for Engineers and Other Scientists, while one-half of the articles read by Mathematicians come from libraries. Computer Scientists are perhaps unique in that they rely about equally on browsing, other people, and printed indexes for identification of articles and also equally on subscription copies, authors, and libraries for article sources.

Table 6.6 completes the picture of the articles read by describing their physical form: actual issue, reprint, preprint, photocopy, and other. In general, these are closely associated with the source of the article. For example, actual issues are either individual or library subscription, and reprints and preprints come either from the author or sometimes directly from the publisher ("other"). Photocopies may be either of library or individual subscription copies or of reprints or preprints. As the table shows, the use of actual issues of journals dominates, accounting for over 80 percent of the total readings. Reprints and preprints together account for an additional 5 percent of readings, while use of other forms (including microforms) is negligible. Photocopies are the source of the remaining 12 percent of readings. The most striking variations on this pattern within the fields of science are higher levels of use of preprints by Computer Scientists, reprints by Environmental Scientists, and photocopies by Mathematicians.

The Purposes of Reading

The general category of purpose for which an article is read provides an indication of its use and serves as an indirect indicator of journal system benefit. Generally we have classified the purposes of reading into six categories:

1. Research: To prepare a proposal or to plan a project, to apply its methodology to a current project, or to apply its findings to a current project.
2. Education (Others): To prepare a lecture or presentation.
3. Education (Self): For professional development, current awareness, or general interest.
4. Writing: To prepare an article, book, review, or report.
5. Citation: To cite in a journal article, book, review, or report.
6. Management: To use for planning, budgeting, or management of research.

The purposes cited by scientists and engineers in our users' survey are shown in Table 6.7.

Purpose frequently includes self-education (94 percent), applying the article's methods to current research (45 percent), applying the article's findings to current research (44 percent), and citation (34 percent). Overall readers identified an average of almost three purposes of use for each article read—self-education and about two other purposes. The high proportion of articles read that are directly applicable to current research is somewhat surprising. The level of citation uses is approximately equal to that for writing, when writing proposals is included. Variations noted among the fields of science include lower levels of use for writing and educating others in the Computer Sciences and Engineering and low levels of application of research findings and methodology in the Environmental and Other Sciences. Scientific journal articles were read for management purposes primarily by Psychologists, Physical Scientists, and Life Scientists.

Table 6.7 Purpose of use, by field of science, 1977 (in percent).

Field of Science	Research Findings	Research Methods	Research Proposal	Writing	Citation	Self-Education	Other's Education	Management
Physical Sciences	66	21	25	31	29	75	46	19
Mathematics	63	50	8	36	45	93	20	9
Computer Sciences	66	66	1	1	1	100	1	[a]
Environmental Sciences	10	18	10	10	13	99	30	4
Engineering	40	60	20	1	40	100	1	[a]
Life Sciences	52	32	19	36	39	91	39	14
Psychology	27	27	15	33	21	100	38	19
Social Sciences	42	26	17	45	41	84	26	7
Other Sciences	12	4	1	15	35	96	54	6
All Fields	44	45	17	16	34	94	18	6

Note: Proportions do not sum to 100 percent since multiple answers were received.
[a]Less than 0.005.

Table 6.8 Purposes of uses for life scientists and cancer researchers, 1977 (in percent).

Purpose of Use	Life Scientists	Cancer Researchers
Research proposal	19	17
Research methods	32	28
Research findings	52	68
Education (others)	39	14
Education (self)	91	62
Writing	36	N.A.
Citation (in journal)[a]	39	23
Citation (other)	a	10
Management	14	N.A.

Source: King Research, Inc. User Survey; D. W. King, D. D. McDonald, and
  C. H. Olsen, *A Survey of Readers, Subscribers, and Authors of the* Journal of the
  National Cancer Institute (Rockville, Maryland: King Research, 1978).
Note: Combined figure for all citations used.
[a]In journal for cancer researchers only.

The only other scientific and technical journal study that investigated uses of information on a nationwide basis is that performed for the National Cancer Institute.[8] Comparisons of these data with the Life Sciences results are given in Table 6.8. The *JNCI* study gives us some evidence of the proportion of all citations to journal articles. We found that 23 percent of the uses of journal articles were for citations in other journal articles and 10 percent for citations in a book, review, or report. Thus, roughly, one might say that about 70 percent of all citations to journal articles occur in other journal articles. Extrapolating to the total number of article uses suggests that the total number of citations to journal articles might be on the order of 58 million per year.

THE RELATION OF CITATION ANALYSES TO JOURNAL USE DATA

Citation counts are commonly used as a surrogate measure of use and value.[9,10,11] They have been used for such purposes as evaluating journals,[12] articles,[13] authors,[14] research funding, and amount of transnational flow of information.[15,16,17] They have also been used to describe the structure of scientific literature[18,19,20] and network theory.[21] Narin and Moll[22] have referred to a number of studies that illustrate the usefulness and validity of applying citation analysis:

K. E. CLARK correlated a variety of bibliometric measures used to select a panel of eminent psychologists and found that citation measures were especially significant in indicating eminence.

BAYER & FOLGER found that citation measures correlate positively with the quality of a scientist's education.

COLE & COLE (1967), sociologists, found that citation data correlated highly with survey data in determining the quality of a scientist's work.

SOLOMON, another sociologist, computed the correlations in studies ranking graduate programs in sociology with a set of productivity indexes compiled from published books and major articles.

BUSH ET AL. compared citation rankings of journals in economics with a study using expert opinions to obtain similar rankings, and found that the two rankings were "remarkably close."

MARTINO, in an article advocating use of citation data for research and development (R&D) management, reviewed a number of studies correlating citation frequency with quality of research and reported a study of work evaluation by the Air Force Office of Scientific Research. He concluded that citation patterns correlate with the quality of work.

VIRGO compared peer review evaluation of pairs of scientific papers with an evaluation of the same papers using citation data. She found that citation data was able to consistently mirror the judges' opinions.

COHEN-SHANIN found that peer judgments of the innovative quality of research papers in plant physiology correlated very highly with rates of citations to these papers.

The authors concluded, "Despite the mass of evidence validating bibliometric analysis, negative reactions persist, based more on subjective fear and emotion than fact."

   Citation counts provide an excellent surrogate measure for use and/or value in many instances. They also provide a good source for analysis of communication structure and networks. However, there are two major problems concerning use of citations for such analyses: citations appear over a long time following publication, and citation usage takes place over a substantially longer period of time than the other purposes of use. We have observed that the communication paths, as a consequence, differ substantially depending on the purpose of use. Thus if citations are used for evaluation purposes, one should be assured that there is no analytical dependency on age or on the way in which articles are obtained.

   The most common measure of age of articles is its half-life, a term

Table 6.9 Ages of journal articles at time of use (in years).

	Average Age for Citations	Half-Life for Citations	Average Age for All Uses	Half-Life for All Uses
Physical Sciences	9.7		2.3	0.2
Physics		4.6		
Chemistry		8.1		
Mathematics	15.0	10.5		
Computer Sciences	5.9			
Environmental Sciences	8.8			
Geology		11.8		
Engineering	7.4			
Chemical		4.8		
Mechanical		5.2		
Metallurgical		3.9		
Life Sciences	11.2			
Physiology		7.2		
Botany		10.0		0.25
Psychology	10.4			0.9
Social Sciences	11.9			
Other Sciences	12.2			
All Fields	10.8			0.2

Sources: D. W. King, D. D. McDonald, N. K. Roderer, C. G. Schell, C. G. Schueller, and B. L. Wood, *Statistical Indicators of Scientific and Technical Communication (1960–1980),* 2d ed. (Rockville, Maryland: King Research, 1977); B. Houghton, *Scientific Periodicals* (Hamden, Connecticut: Linnet Books, 1975); A. J. Meadows, *Communication in Science* (London: Butterworths, 1974); King Research, Inc. User Survey; Case Institute of Technology, *An Operations Research Study of the Dissemination and Use of Recorded Scientific Information* (Cleveland, Ohio: Case Institute of Technology, 1960); D. W. King, D. D. McDonald, and C. H. Olsen, *A Survey of Readers, Subscribers, and Authors of the* Journal of the National Cancer Institute (Rockville, Maryland: King Research, Inc., 1978), American Psychological Association, *Reports of: The American Psychological Association's Project on Scientific Information Exchange in Psychology,* Vol. 1 (Washington, D.C.: American Psychological Association, 1963).

borrowed from nuclear physics used to describe nuclear decay over time. Citations resemble such decay to the extent that most citations occur shortly after publication, but some may extend over a long period of time. The half-life of citations is taken as the median age, when one-half the citations to an article have occurred. The average age of cited articles is usually greater than the half-life, due to the skewed shape of the distribution. We observed that the shape of the age curve for frequently cited articles is about the same as the shape of infrequently cited articles.[23] Examples of the age distributions of all uses and citations are given in Table 6.9.

Table 6.10 Proportion of readings used for citations and for all uses, 1977 (in percent).

	Readings for All Uses	Readings for Citations
Source of article		
Personal subscription	68	12
Library subscription or interlibrary loan	15	67
Author	12	13
Coworker	4	8
Form of article		
Actual issue	83	65
Reprint	3	13
Preprint	2	3
Microform	a	1
Photocopy	11	18
Paper copy from microform	a	0.3

Source: King Research, Inc. User Survey; King Research, Inc.
[a]Less than 0.005.

The differences observed in the sources and forms of articles used for citation purposes, compared to all purposes, are shown in Table 6.10.

Time Spent Reading

One indication of the value of journals is the amount of time that scientists and engineers are willing to spend reading them. Reading time is also the major cost element in determining journal system costs. Several studies have indicated the amount of time that scientists and engineers spend reading journals. These data are summarized in Table 6.11 by field.

From the data shown, a general pattern emerges, with low levels of reading by Engineers and higher levels for Psychologists, Physical Scientists, and Life Scientists. More complete and comparable data were obtained in our user survey, which asked directly about the amount of time spent reading scientific journal articles in the last month. The results (Table 6.12) compare well with previous results and indicate some new information—chiefly high levels of reading time for Mathematicians and Social Scientists.

The amount of time spent reading can also be compared with the number of articles read, which provides another indication of level of

Table 6.11   Time spent by scientists and engineers reading
journals, by field of science, 1960–1977.

Field of Science	Hours per Month Reading Journals	Hours per Month Reading All Literature
Physical Sciences	9.0	
	25.1	36.8
	11.7	24.3
	4.8–11.7	
Engineers	5.0	
	19.1	45.1
	2.2–3.5	8.6–13.8
Life Sciences (cancer)	8.1	
	24.1	
Psychology	4.8	
	13.9[a]	27.7
All Fields	6.8	
	7.8–10.4	15.6–20.8
	11.7[a]	23.4

Sources: King Research, Inc. User Survey; American Chemical Society: Case
  Institute of Technology, *An Operations Research Study of the Dissemination and
  Use of Recorded Scientific Information* (Cleveland, Ohio: Case Institute of
  Technology, 1960); B. Weil, "Benefits from Researcher Use of the Published
  Literature at the Exxon Research Center," Paper presented at the National
  Information Conference and Exposition, Washington, D.C., April 20, 1977;
  T. D. Allen, *Managing the Flow of Scientific and Technological Information*
  (Cambridge, Massachusetts: Massachusetts Institute of Technology, 1966);
  D. W. King, D. D. McDonald, and C. H. Olsen, *A Survey of Readers, Subscribers,
  and Authors of the* Journal of the National Cancer Institute (Rockville, Maryland:
  King Research, 1978); American Psychological Association, *Reports of:
  The American Psychological Association's Project on Scientific Information
  Exchange in Psychology,* Vol. 1 (Washington, D.C.: American Psychological
  Association, 1963); D. W. King and N. K. Roderer, *Systems Analysis of Scientific
  and Technical Communication in the United States: The Electronic Alternative to
  Communication Through Paper-Based Journals* (Rockville, Maryland: King
  Research, 1978); A. M. Hall, P. Clague, and T. M. Aitchison, *The Effect of the Use
  of an SDI Service in the Information-Gathering Habits of Scientists and
  Technologists* (London: Institute of Electrical Engineers, 1972), p. 000.
[a]Computed as one-half of total literature.

reading (Table 6.13). The average time spent reading an article ranges
from thirty minutes in several fields to nearly four hours in Mathematics.
Related data on the level and purpose of reading (see Tables 6.3 and
6.7) do not provide any direct justification for the amount of reading
time spent by Mathematicians; in fact, they indicate that a higher
proportion of articles are read just to get the idea and for a variety of
purposes. One explanation might be the length of Mathematics articles,
which at 10.2 pages are considerably longer than the overall average

Table 6.12   Time spent by scientists and engineers
reading journals, 1977.

Field of Science	Hours per Month Spent Reading Journals
Physical Sciences	9.0
Mathematics	19.1
Computer Sciences	3.4
Environmental Sciences	4.2
Engineering	5.0
Life Sciences	8.1
Psychology	4.8
Social Sciences	11.3
Other Sciences	7.2
All Fields	6.8

Table 6.13   Average time spent per article read, by field of science,
1977.

Field of Science	Hours per Month Spent Reading	No. of Articles Read Monthly	Minutes per Article
Physical Sciences	9.0	15.8	35
Mathematics	19.1	5.1	225
Computer Sciences	3.4	5.6	35
Environmental Sciences	4.2	9.0	30
Engineering	5.0	6.4	45
Life Sciences	8.1	15.0	30
Psychology	4.8	8.9	30
Social Sciences	11.3	10.4	65
Other Sciences	7.2	6.3	70
All Fields	6.8	8.7	45

length (7.5 pages). Other areas where reading time is high are Social
Sciences and Other Sciences. In the Social Sciences again we find a
relatively low level of intensity in reading, many purposes of use, and
lengthy articles (13.3 pages). In Other Sciences, uses mentioned are the
more general, perhaps time-consuming ones.

Other Assessments of Benefits of Scientific and
Technical Journals

It is very hard to measure the actual value derived from scientific
and technical journals. For a rough estimate, we asked respondents to

Table 6.14   Savings due to journal reading, by field of science, 1977.

Field of Science	Readings in Which Time Is Saved (%)	Total Value of Time Saved (millions)	Total Value of Other Savings (millions)
Physical Sciences	6	$0.2	
Mathematics	22		
Computer Sciences	33	0.1	
Environmental Sciences	10	0.1	
Engineering	33	53.2	$52.9
Life Sciences	22	1.5	0.9
Psychology	14	4.1	
Social Sciences	15	0.5	
Other Sciences	7	0.1	
All Fields	24	59.7	53.8

indicate (1) whether reading a specific article saved any time on a current task or project; (2) to provide the dollar value (direct salaries only) of any time saved; and (3) to provide the dollar value (nonsalary items) of any time saved. The results of the survey are summarized in Table 6.14, which shows that almost one-quarter of the respondents overall were able to identify time savings as a result of reading the specified journal article. An additional 20 percent were uncertain as to whether any time had been saved. Time savings occurred most frequently in Computer Sciences and Engineering, with savings in Engineering (expressed in dollars) being far greater than those in other fields.

COST OF USE OF SCIENTIFIC AND TECHNICAL JOURNALS

The cost of journal use depends largely on the amount of time spent by users finding and reading scientific and technical articles. In our cost model, we incorporate several activities, including individual searching, data-base searching, photocopying, and reading.

Individual Acquisition of Subscriptions

Input: Scientist's decision to subscribe to a journal.
Process: Placement of order or renewal and payment of

subscription price.
Output: A paid individual subscription.

Whether initiating a new subscription or renewing an already existing one, the scientist or an agency must notify the publisher and remit the subscription price. This is generally a one-step operation, estimated to take fifteen minutes of a support person's time. It is generally performed on an annual basis for each individual subscription.

Individual Acquisition of Separates

Input: Scientist's awareness of and need for a preprint or reprint of a journal article.
Process: Transmittal of request to author or publisher.
Output: A completed preprint or reprint request.

Preprints are almost always obtained directly from the author; reprints may be requested from the author, the publisher, or a journal reprint service. Requests may be transmitted orally or in writing, and they might not require payment for the article copy. It is assumed that, on the average, a request for a reprint or preprint will take fifteen minutes of a support person's time. The number of requests for reprints and preprints is assumed to be the same as the number of reprint and preprint copies produced.

Individual Storage of Subscriptions

Input: Subscription copies received by scientist.
Process: Retention of journal for possible later use.
Output: Stored journal volume.

The storage of journal subscription copies by scientists is generally a straightforward matter of keeping a run of journal issues on a convenient bookshelf. Since it is estimated (for 1977) that physical library storage of a journal volume costs $0.30,[24] the same figure is used for storage of an individual subscription volume.

For all storage costs incurred in a given year to be identified, volumes acquired in previous years and still retained must be taken into account. This can be done by assuming that journals will be stored for fifteen years on the average and then summing the number of individual subscription copies acquired over that time period. The result of one such calculation is that the number of volumes stored in 1977 is 10.65 times the number of volumes currently acquired. The cost of occasionally weeding out and disposing of journals no longer wanted is ignored.

Individual Storage of Separates

Input: Preprints and reprints received by scientist.
Process: Retention of preprints and reprints for possible later use.
Output: Stored preprints and reprints.

Storage of preprints and reprints is similar to that of individual subscription copies, except that they are likely to be placed in vertical files and might be organized by category. It is assumed that copies will be stored over a fifteen-year period. Accounting for all reprints and preprints acquired between 1963 and 1977, we find that in 1977 the number of reprints and preprints stored was 11.55 times the number currently acquired. Again the cost of weeding is ignored.

Individual Searching

Input: An information need.
Process: Identifying and/or accessing journal article.
Output: Journal article available for reading.

Individual searching involves the activities performed by the scientist prior to reading a particular journal article. In general, we assume that uses of articles published within the past year generally involve journals that the scientist subscribes to or regularly scans,* and thus do not involve extensive searching. This small amount of time is subsumed in reading time. Uses of articles over one year old may be preceded by discussions with colleagues, use of a printed index, use of a machine-readable data base, tracing citations in more recent primary publications, and so on and may also require additional action to acquire the article, such as through a visit to a library. We assume that the average time spent searching for articles more than one year old is ten minutes. The proportion of uses of articles more than one year old is 35 percent over all the fields of science.

*One significant exception to this pattern is the time spent on use of selective dissemination of information (SDI) and other current awareness services.

Data-Base Searching

This is a special case of individual searching and may involve the services of an intermediary.

Input: An information need.
Process: Formulation of the information request according to the data-base service protocol and actual conduct of searches.
Output: List of potentially useful articles, with or without abstracts.

Costs for automated searching are incurred for the time to formulate queries, time spent by personnel at the on-line terminal, computer time, communication equipment and lines, and (for massive printouts) supplies and materials handling. Costs for making the data bases available to be searched were discussed in Chapter 5. We estimate that the total costs associated with data-base searching as such in 1977 were $107 million.[25] This is equivalent to an average cost of about $75.

Photocopying

Input: Journal article.
Process: Photocopying of the article.
Output: Photocopy.

Frequently articles are photocopied before they are used. This applies particularly to library subscription copies of an article, where the original often cannot be taken from the library. In the user survey, 10 percent of articles obtained by users were photocopied. Average photocopying costs in 1977 were estimated at $0.21 per page or $1.70 per article.

Reading

Input: Article available for use.
Process: Scientist reads past the title and abstract.
Output: Increase in stored knowledge; one additional use of article.

The cost of reading journal articles was derived from estimates of the amount of time spent reading. For 1977, it was estimated that the average scientist spends about 1.6 hours per week reading journals, or 3 percent of total working time (assumed to average forty-eight hours per week). Combining this figure with the median salary of scientists and engineers, we can find an annual cost per scientist (Table 6.15).

Overall Costs of Journal Use

Combining the several costs for the activities involved in use of journal articles, and carrying over the various assumptions made, overall costs can be developed. These are presented by field of science in Table 6.16.

FEDERAL CONTRIBUTION FUNDING TO USE OF SCIENTIFIC AND TECHNICAL JOURNALS

Nearly 20 percent of the scientists and engineers included in our user survey were employed by the federal government. An additional 10 percent indicated that the government was their primary source of funding. Table 6.17 indicates the number of federally supported scientists by field of science, along with their level of reading and the associated reading costs. Of the $2.4 billion total associated with the reading of journal articles, nearly $900 million (37 percent) can be estimated to come from federal funds directly or indirectly. Because other use-related costs are closely tied to reading costs, we estimate that the same proportion of acquisition, storage, searching, data-base searching, and photocopying cost are federally supported.

REFERENCES

1. American Psychological Association, *Reports of: The American Psychological Association's Project on Scientific Information Exchange in Psychology,* Vol. 1 (Washington, D.C.: American Psychological Association, 1963).
2. Case Institute of Technology, *An Operations Research Study of the Dissemination and Use of Recorded Scientific Information* (Cleveland, Ohio: Case Institute of Technology, 1960).
3. F. Machlup and K. W. Leeson, *Information Through the Printed Word: The Dissemination of Scholarly, Scientific and*

Table 6.15 Estimated user costs associated with reading journals, by field of science, 1977.

Field of Science	No. of Active Scientists and Engineers (000)	Average Annual Reading Time (hours)	Median Annual Salaries (000)	Annual Reading Cost per Scientist or Engineer (000)	Total Reading Cost (millions)
Physical Sciences	207.8	96	$24.2	$1.40	$285.3
Mathematics	97.3	175	24.6	2.60	252.5
Computer Sciences	161.4	35	22.9	0.50	74.9
Environmental Sciences	75.5	50	25.0	0.70	55.4
Engineering	1,082.3	53	24.1	0.70	764.4
Life Sciences	276.3	93	22.1	1.20	329.5
Psychology	107.3	55	24.1	0.80	83.1
Social Sciences	226.2	125	25.1	1.80	409.2
Other Sciences	99.9	48	N.A.	0.70	66.8
All Fields	2,334.4	72	24.0	1.00	2,354.1

Table 6.16 Costs associated with reading of journal articles, by field of science, 1977.

Field of Science	Cost Components (millions)								Total Cost (millions)	No. of Active Scientists and Engineers (000)	Average Annual Cost per Scientist and Engineer
	Individual Acquisition of Subscription	Individual Acquisition of Separates	Individual Storage of Subscription	Individual Storage of Separates	Individual Searching	Data-Base Searching	Photocopying	Reading[a]			
Physical Sciences	$0.6	$12.6	$1.0	$0.3	$33.3	$17.3	$8.1	$285.3	$358.6	207.8	$1.70
Mathematics	.5	.8	.8	a	8.4	2.6	4.3	252.5	270.0	97.3	2.80
Computer Sciences	.9	1.2	1.4	a	8.2	4.8	a	74.9	91.5	161.4	0.60
Environmental Sciences	3.4	6.6	5.5	.1	6.8	3.6	1.8	55.4	83.3	75.5	1.10
Engineering	3.2	7.8	5.2	.2	65.7	36.4	.1	764.4	883.0	1,082.3	0.80
Life Sciences	11.1	56.2	18.0	1.2	37.3	21.9	19.1	329.5	494.5	276.7	1.80
Psychology	.5	3.4	.8	.1	5.0	5.1	3.1	83.1	101.2	107.3	0.90
Social Sciences	3.4	7.3	5.5	.2	34.8	12.5	18.9	409.2	491.7	226.2	2.20
Other Sciences	7.4	8.8	11.9	.2	6.2	3.3	1.4	66.8	106.0	99.9	1.10
All Fields	31.2	106.4	50.3	2.4	205.0	107.4	46.6	2,354.1	2,903.3	2,334.4	1.20

Note: Only costs incurred by individual user are included here. Not included are costs such as subscription fees, cost of maintaining a data base of library, or most transmittal costs.

[a] Less than one-half of one-tenth of one percent.

Table 6.17 Estimated user costs associated with reading journals, by federally supported scientists and engineers, by field of science, 1977.

Field of Science	No. of Federally Supported Scientists and Engineers (000)	Average Annual Reading Time (hours)	Annual Reading Costs per Scientist and Engineer (000)	Total Reading Cost (millions)
Physical Sciences	79.0	106	$1.5	$121.8
Mathematics	15.0	229	3.4	50.8
Computer Sciences	1.1	56	0.8	0.8
Environmental Sciences	49.8	32	0.5	23.9
Engineering	280.3	59	0.9	239.5
Life Sciences	139.1	99	1.3	182.9
Psychology	56.4	55	0.8	44.9
Social Sciences	73.9	141	2.1	157.2
Other Sciences	27.9	85	1.2	34.6
All Fields	722.4	83	1.2	864.8

Intellectual Knowledge, vol. I: *Book Publishing,* vol. II: *Journals,* vol. III: *Libraries* (New York: Praeger, 1978).

4. D. W. King, D. D. McDonald, and C. H. Olsen, *A Survey of Readers, Subscribers, and Authors of the* Journal of the National Cancer Institute (Rockville, Maryland: King Research, 1978).
5. D. W. King and N. K. Roderer, *Systems Analysis of Scientific and Technical Communication in the United States: The Electronic Alternative to Communication Through Paper-Based Journals* (Rockville, Maryland: King Research, 1978).
6. Case Institute of Technology, *An Operations Research Study.*
7. Machlup and Leeson, *Information Through the Printed Word.*
8. King, McDonald, and Olsen, *A Survey of Readers.*
9. E. Garfield, "Citation Analysis as a Tool in Journal Evaluation," *Science* 178:471–479 (1972).
10. F. Narin and J. L. Moll, "Bibliometrics," in *Annual Review of Information Science and Technology,* vol. 12, ed. M. E. Williams, (White Plains, New York: Knowledge Industry Publications, 1977).
11. D. S. Price, "A General Theory of Bibliometric and Other Cumulative Advantage Processes," *Journal of the American Society for Information Science* 27(5):292–306 (1976).
12. Garfield, "Citation Analysis as a Tool."
13. E. Garfield, "Citation Indexes for Science," *Science* 122(3159):108–111 (1955).
14. D. S. Price, "Studies in Scientometrics. II. The Relation Between Source Author and Cited Author Populations," *International Forum on Information and Documentation* 1(3):19–22 (1976).
15. J. D. Frame and F. Narin, "NIH Funding and Biomedical Publication Output," *Federal Proceedings* 35(14):2529–2532 (1976).
16. F. Narin, *Evaluative Bibliometrics: The Use of Publication and Citation Analysis in the Evaluation of Scientific Activity* (Cherry Hill, New Jersey: Computer Horizons, 1976).
17. National Science Board, National Science Foundation, *Science Indicators* (Washington, D.C.: Government Printing Office, 1972 and 1974).
18. E. Garfield, "Citation Indexing for Studying Science," *Nature* 227(5259): 669–671 (1970).
19. H. Small and B. C. Griffith, "The Structure of Scientific Literature. I. Identifying and Graphing Specialties," *Science Studies* 4(1):17–40 (1974).
20. H. Small, "Co-Citation in the Scientific Literature: A New Measure of the Relationship Between Two Documents," *Journal of the American Society for Information Science* 24(4):265–269 (1973).

21. B. C. Griffith, H. G. Small, J. A. Stonehill, and S. Dey, "The Structure of Scientific Literature. II. Toward a Macro- and Microstructure for Science," *Science Studies* 4(4):339–365 (1974).
22. Narin and Moll, "Bibliometrics."
23. D. W. King, F. W. Lancaster, D. D. McDonald, N. K. Roderer, and B. L. Wood., *Statistical Indicators of Scientific and Technical Communication,* vol. II: *A Research Report* (Rockville, Maryland: King Research, 1976).
24. V. E. Palmour, M. C. Bellassai, and R. R. V. Wiederkehr, *Costs of Owning, Borrowing, and Disposing of Periodical Publications* (Arlington, Virginia: Public Research Institute, 1977).
25. King et al., *Statistical Indicators.*

Chapter 7

The Flow of Information Through Scientific and Technical Journals

THE FLOW OF JOURNALS

The simplest way to describe the flow of information from journal authors to users is by the diagram shown in Figure 1.5 in Chapter 1 and repeated here as Figure 7.1. There are four principal groups of participants in the flow of journal information: scientists as authors and users, publishers, secondary organizations, libraries. Each group represents several other participants. For example, authorship includes secretarial, internal editing, and graphics support. Publishers as a participant group includes reviewers, editors, and sometimes outside typesetters, printers, and brokers. Scientists serve both as end users and as intermediaries. Often a scientist will learn about or obtain a journal article from a colleague.

In the flow diagram there are fourteen principal paths or links of communication among the participants. The principal communication unit is referred to here as a message, which could be a manuscript, article, abstract, or even a letter of transmittal about an article or journal. In order to establish the order of magnitude concerning the extent of communication, the estimated number of messages that are transmitted is summarized in Table 7.1. In the table each major type of communication is described by the path number (1 to 14), the principal sending participant, the message form or medium (for example, manuscript or article), the channel (such as mail or personal delivery), the principal receiving participant, and the estimated number of messages transmitted in 1977.

The initial communication in the flow involves the correspondence between authors and coauthors who may be colleagues or may work in different organizations. Since there is an average of almost two coauthors per article, it is assumed that a draft manuscript passes between them at least once. There were an estimated total of 382,000

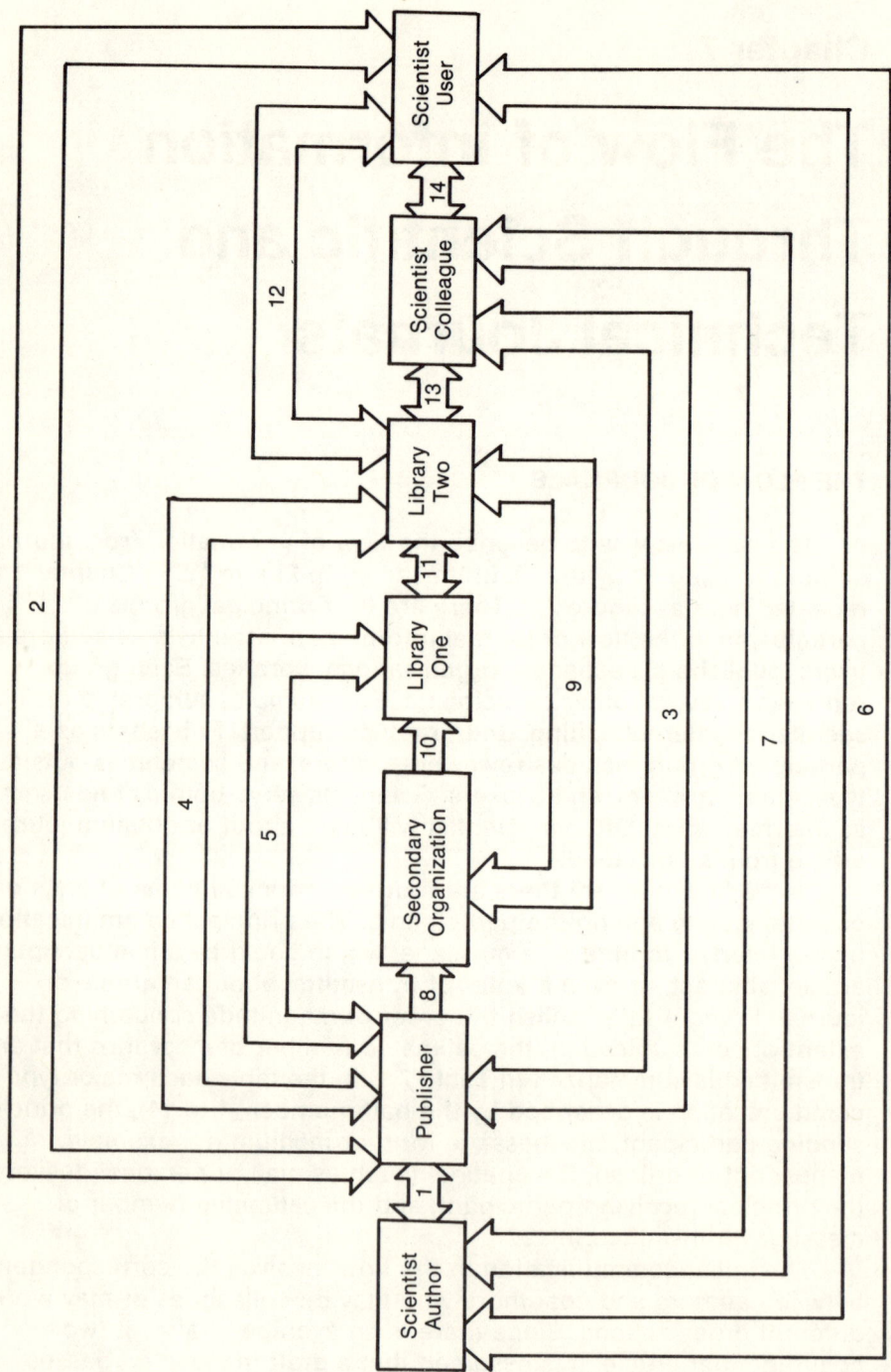

Figure 7.1 General flow of information among scientists and engineers.

Table 7.1 Estimated number of messages transmitted from sender to receiver, by channel, 1977.

Path Number	Sending Participant	Message Form (Medium)	Channel	Receiving Participant	Number of Messages Transmitted
Authors	Scientist (author)	Draft manuscript	Mail/personal	Scientist (coauthor)	399,000 draft manuscripts
Authors	Scientist (coauthor)	Draft MSS/letter	Mail/personal	Scientist (author)	399,000 draft MSS/letters
7	Scientist (author)	Draft manuscript	mail/personal	Scientist (colleague)	2 million draft manuscripts
7	Scientist (colleague)	Draft MSS/letter	Mail/personal	Scientist (author)	2 million draft manuscripts
1	Scientist (author)	Draft manuscript	Mail	Publisher	778 million draft manuscripts
1	Publisher	Draft MSS/letter	Mail/personal	Scientist (author)	778 million draft MSS/Letters
3	Publisher	Draft manuscript	Mail	Scientist (reviewer)	876,000 draft manuscripts
3	Scientist (reviewer)	Draft MSS/letter	Mail	Publisher	876,000 reviews
2, 3	Scientist (user or colleague)	Letter	Mail	Publisher	15.8 million subscription requests
2, 3	Publisher	Journal issues	Mail	Scientist (user or colleague)	104 million journal issues
4, 5	Library	Letter	Mail	Publisher	8.0 million subscription requests
4, 5	Publisher	Journal issues	Mail	Library	5.3 million journal issues
4, 5	Publisher	Reprints	Mail	Library	Unknown
8	Secondary organization	Letter	Mail	Publisher	13,000 letter requests

189

Table 7.1 *(continued)*

Path Number	Sending Participant	Message Form (Medium)	Channel	Receiving Participant	Number of Messages Transmitted
8	Publisher	Journal issues/reprints	Mail	Secondary organization	88,000 journal issues
9, 10	Library	Letter	Mail	Secondary organization	Unknown
9, 10	Secondary organization	A&I publications	Mail	Library	Unknown
9	Secondary organization	Searches	Telecommunication	Library	1.4 million searches
12, 13	Library	Article (journal issue)	Personal/mail	Scientist (user or colleague)	28.3 million articles
12, 13	Library	Article (photocopy)	Personal/mail	Scientist (user or colleague)	7.5 million article photocopies
12, 13	Library	Article (microform)	Personal/mail	Scientist (user or colleague)	52,000 article microforms
12, 13	Library	Article (reprint)	Personal/mail	Scientist (user or colleague)	288,000 article reprints
11	Library (2)	Letter/ILL form article	Mail/telecommunication	Library (1)	4.0 million ILL requests
11	Library (1)	Article photocopies	Mail	Library (2)	4.0 million ILL fulfillments
1	Publishers	Reprints	Mail	Scientist (author)	53 million reprints

	Sender	Item	Method	Recipient	Quantity
2, 3	Scientist (user or colleague)	Letter	Mail	Publisher	600,000 reprint requests
2, 3	Publisher	Article reprints	Mail	Scientist (user or colleague)	600,000 reprints
6, 7	Scientist (user or colleague)	Letter	Mail	Scientist (author)	800,000 preprint requests
6, 7	Scientist (author)	Article preprint	Mail	Scientist (user or colleague)	2 million preprints
6, 7	Scientist (user or colleague)	Letter	Mail	Scientist (author)	26 million reprint requests
6, 7	Scientist (author)	Article reprint	Mail	Scientist (user or colleague)	26 million reprints
6, 7	Scientist (user or colleague)	Letter	Mail	Scientist (author)	1.5 million requests
6, 7	Scientist (author)	Article photocopies	Mail	Scientist (user or colleague)	1.5 million photocopies
14	Scientist (colleague)	Article (journal issues)	Personal	Scientist (user)	5.8 million articles
14	Scientist (colleague)	Article preprint	Personal	Scientist (user)	84,000 article preprints
14	Scientist (colleague)	Article reprint	Personal	Scientist (user)	300,000 article reprints
14	Scientist (colleague)	Article photocopy	Personal	Scientist (user)	3.5 million article photocopies
Users	Scientist (user)	Article (journal issues)	Personal	Scientist (user)	166 million articles
Users	Scientist (user)	Article photocopy	Personal	Scientist (user)	25 million personal photocopies

articles published in 1977 and an additional 17,000 manuscripts that were rejected and dropped from further action (based on the author survey). Of the articles published, about 39,000 were rejected by the first journal to which they were submitted. It is estimated that there were a total of 399,000 draft manuscripts that were transmitted each way among the authors and coauthors. In addition, according to Garvey et al.[1,2] there are about an average of five preprints per manuscript that are sent to peers for informal review of the manuscript (path 7). Thus we estimate that about 2 million draft manuscripts were sent to peers and that there was another similar number of returned manuscripts, letters, or personal telephone responses.

The next major communication takes place between authors and publishers (path 1). Including the number of articles published, resubmitted, to new journals, and rejected, an estimated 438,000 manuscripts are involved. From the author survey, it is estimated that articles are returned to the author for revisions on the average about once. Thus a total of 778 manuscripts per year are transmitted from authors to publishers and 778 manuscripts (or letters) transmitted from publishers to authors. In addition, it is estimated that 438 article manuscripts are reviewed by an average of two reviewers each. Thus a total of 876 manuscripts are sent to reviewers, and a similar number of reviews were returned per year (path 3).

When an article is published, it is distributed to subscribers. There are estimated to be 15.8 million individual (or personal) subscriptions per year distributed by publishers in 1977. Thus it is assumed that there was this number of requests for subscriptions. There was an average of 6.6 issues per journal so the total number of issues sent from publishers to individuals was 104 million. This represents an astounding 1.4 billion copies of articles per year that were transmitted by publishers to individual U.S. scientists.

Similarly there was an estimated total of 8.0 million institutional (library) subscriptions (paths 4,5). The total number of issues transmitted to libraries was about 53 million. The number of copies of articles transmitted to institutions was 700 million. Publishers also distributed a certain number of reprints to libraries, but this number is unknown. Publishers send copies of their journals to secondary organizations (path 8) that abstract and index journal articles. It is estimated that the average coverage of journals was three secondary organizations per journal (based on *Ulrich's*). Thus 88,000 issues were transmitted from publishers to secondary organizations.

The secondary organizations, in turn, send the bibliographic publications to libraries. Unfortunately the number of transmissions is unknown. It is estimated, however, that a total of 1.4 million searches were performed using computers. Most of these transmissions are on-line using telecommunication. Also some secondary services distribute reprints or copies on request to libraries. For example, the Institute for Scientific Information, Inc. distributed about 400,000 copies through their original article tear sheet (OATS) service. In the absence of additional data, this figure is used.

Libraries make journal articles available to scientists and engineers in a number of forms (paths 12,13). The number of readings is estimated from the 1977 user survey with the following results: journal paper-copy issues, 28.3 million; photocopies of articles, 7.5 million; journal microform or microform hard copy, 52,000; and article reprint, 288,000. In most instances the articles have been made available through circulating copies or in the library. Also it is estimated[3] that there are 4.0 million copies made available to scientists through interlibrary loans (path 11). Requests from one library to another are usually transmitted by letter but sometimes by telecommunication. Most of the responses are in the form of mailed photocopies.

Publishers often distribute copies of reprints to authors, who in turn transmit them to users. Based on an earlier author survey, it is estimated that the average number of reprints received by authors is 138 and the total number of reprints received by authors is 53 million (path 1). Users may also request and receive reprints directly from publishers (paths 2,3). It is estimated that this number is about 600,000. The number of preprints transmitted by authors on request to scientists (paths 6,7) is estimated to be 800,000. Similarly the number of reprints distributed by authors to scientists (paths 6,7) is estimated at 26 million. These reprint transmissions are initiated by both the author and user on request. Finally authors may also distribute photocopies of their articles to scientists in their field either directly or on demand (paths 6,7). The estimated number of photocopies sent from authors to other scientists is 2 million preprint copies plus 1.5 million other copies.

Many copies of journal articles are exchanged among colleagues who are coworkers (path 14). It is estimated from the user survey that these transmissions involved the following message forms (media) to the degree specified: journal paper-copy issues, 5.8 million; preprints, 84,000; reprints, 300,000; and photocopies, 3.5 million. Clearly colleagues form an important link in the communication flow.

Many of the readings by scientists take place from their own copies. We estimate this number to have been about 166 million. Of this number it was observed in the user survey that users made 25 million photocopies of articles for their own use.

Summing the total number of transmissions made in one form or another, we find that there were an estimated 2.1 billion such

Table 7.2 Estimated number of readings of scientific and technical journals, by source and field of science, 1977 (in millions of readings).

Field of Science	Publishers	Libraries	Authors	Total Readings
Physical Sciences	30.2	8.8	0.3	39.3
Mathematics	2.5	2.9	0.5	5.9
Computer Sciences	3.8	3.6	3.5	10.9
Environmental Sciences	5.4	1.3	1.5	8.2
Engineering	68.1	0.6	13.9	82.6
Life Sciences	33.0	13.1	3.6	49.7
Psychology	8.6	2.9	0.1	11.5
Social Sciences	19.9	6.8	1.6	28.3
Other Sciences	6.8	0.6	0.2	7.6
All Fields	178.6	36.1	29.5	244.0

transmittal messages in 1977. The most clearly distinguishable paths involve readings from the following sources:

1. Publishers: Journals or reprints sent from publishers to scientists (users or colleagues).
2. Libraries: Journal copies, reprints, or interlibrary loans transmitted from libraries to scientists (users or colleagues).
3. Authors: Preprints, reprints, or photocopies transmitted from authors to scientists (users or colleagues).

The number of readings involving each of these three aggregated paths is given in Table 7.2.

Readings from Copies Transmitted from Publishers to Scientists

By far the greatest proportion of readings comes from copies of articles that are transmitted directly from publishers in the form of journal subscriptions or reprints. Availability of information from this source is described in Table 7.3. The average use of an article (over its lifetime) is about 470, ranging from 150 in the Other Sciences to 1,461 in Engineering. On the average, about one of every eight copies distributed is used. The highest levels of use per copy are in Engineering, the Physical Sciences, and Psychology, with about one use for every two copies distributed.

Table 7.3 Availability and readings of articles transmitted directly from publisher to scientists, by field of science, 1977.

Field of Science	Articles Written (000)	Readings (millions)	Readings per Article	Article Copies Distributed (millions)	Readings per Article Copy Distributed
Physical Sciences	57.5	30.2	525	65.3	0.46
Mathematics	8.6	2.5	291	21.3	0.12
Computer Sciences	9.0	3.8	422	61.4	0.06
Environmental Sciences	17.0	5.4	318	120.7	0.04
Engineering	46.6	68.1	1,461	160.4	0.42
Life Sciences	127.8	33.0	258	552.6	0.06
Psychology	13.4	8.6	642	16.2	0.53
Social Sciences	57.1	19.9	349	68.9	0.29
Other Sciences	45.2	6.8	150	518.2	0.01
All Fields	382.3	178.6	467	1,354.0	0.13

Table 7.4 Availability and readings of articles transmitted directly from libraries to scientists, by field of science, 1977.

Field of Science	Articles Written (000)	Readings (millions)	Readings per Article	Article Copies Distributed (millions)	Readings per Article Copy Distributed
Physical Sciences	57.5	8.8	153	68.2	0.13
Mathematics	8.6	2.9	337	14.0	0.21
Computer Sciences	9.0	3.6	400	66.1	0.05
Environmental Sciences	17.0	1.3	76	51.9	0.03
Engineering	46.6	0.6	13	132.1	0.005
Life Sciences	127.8	13.1	103	185.0	0.07
Psychology	13.4	2.9	216	23.2	0.13
Social Sciences	57.1	6.8	119	47.8	0.14
Other Sciences	45.2	0.6	13	195.9	0.003
All Fields	382.3	36.1	94	683.9	0.05

Readings from Copies Transmitted from Libraries to Scientists

The second largest proportion of readings comes from copies of articles that are obtained by scientists from libraries. These include readings from journal issues (paperform or microform), reprints, and interlibrary loans. Availability of this source is described in Table 7.4. Both uses per article and uses per article copy are considerably lower than for individual subscriptions, being ninety-four uses per article and one use for every twenty article copies, respectively. The highest level of use per copy is in Mathematics, at one use for every five copies.

Table 7.5 Availability and readings of articles transmitted directly from authors to scientists, by field of science, 1977.

Field of Science	Articles Written (000)	Readings (millions)	Readings per Article	Article Copies Distributed (millions)	Readings per Article Copy Distributed
Physical Sciences	57.5	0.3	5	6.2	0.05
Mathematics	8.6	0.5	58	0.4	1.25
Computer Sciences	9.0	3.5	389	0.6	5.83
Environmental Sciences	17.0	1.5	88	3.2	0.47
Engineering	46.6	13.9	298	3.9	3.56
Life Sciences	127.8	3.6	28	28.8	0.13
Psychology	13.4	0.1	7	1.8	0.06
Social Sciences	57.1	1.6	28	3.7	0.43
Other Sciences	45.2	0.2	4	4.3	0.05
All Fields	382.3	29.5	77	54.7	0.54

Readings from Copies Transmitted from Authors to Scientists

The third aggregated set of paths consists of the copies of articles transmitted directly from authors to scientists. These include author preprints, reprints, and photocopies. Availability of this source is given in Table 7.5. The total number of uses is low, but the average uses per copy is 0.54. In Mathematics, Engineering, and the Computer Sciences we estimate more than one use per copy. Per-copy uses of reprints and preprints is particularly low in Psychology and the Physical and Other Sciences.

Comparing Tables 7.3 through 7.5, we can observe the relative preferences of the fields of science for the three paths, with respect to obtaining articles and to reading them and to the relationships between the two activities. In the Social Sciences, for example, 57 percent of the copies distributed are in the form of individual subscriptions, but this path serves as the source of 70 percent of the uses. This is reflected in a high use per copy figure. Library copies, on the other hand, account for 40 percent of the copies distributed but only 24 percent of the uses.

The availability of article copies across all three paths is given in Table 7.6. Overall there are an estimated 638 uses per article and one use for every eight article copies. Total uses are high in Engineering, due to heavy use of individual subscription and reprint copies, and in Computer Sciences, due to heavy use of all three article forms. The lowest levels of both use and per copy use are found in the Social Sciences. The highest level of per-copy use is one out of every four copies overall, found in the Physical Sciences and Psychology.

Table 7.6 Availability and readings of article copies from all sources, by field of science, 1977.

Field of Science	Articles Written (000)	Readings (millions)	Readings per Article	Article Copies Distributed (millions)	Readings per Article Copy Distributed
Physical Sciences	57.5	39.3	683	139.7	0.28
Mathematics	8.6	5.9	686	35.7	0.17
Computer Sciences	9.0	10.9	1,211	128.1	0.09
Environmental Sciences	17.0	8.2	482	175.8	0.05
Engineering	46.6	82.6	1,773	296.4	0.28
Life Sciences	127.8	49.7	389	766.4	0.06
Psychology	13.4	11.5	858	41.2	0.28
Social Sciences	57.1	28.3	496	120.4	0.24
Other Sciences	45.2	7.6	168	718.4	0.01
All Fields	382.3	244.0	638	2,092.6[a]	0.12

[a]These copies are distributed to U.S. subscribers only. With non-U.S. subscribers added, the total would be about 2.5 billion article copies distributed.

The Communication Paths Associated with Small, Medium, and Large Journals

There is considerable evidence to suggest that the size of a particular journal (in terms of annual circulation) is an important determinant of the cost of producing and using the journal and that, in addition, the activities associated with communicating may vary depending on journal size.

In order to simplify analysis, the universe of scientific and technical journals has been subdivided into three size categories—small, medium, and large, corresponding to total circulation counts of less than 3,000, 3,000 to 10,000, and greater than 10,000, respectively. The distribution shows that over 50 percent of the journals considered were small, about 36 percent medium, and the remaining 13 percent large. Because of the highly skewed distribution, the three classes account for 11, 31, and 58 percent of the total journal circulation, respectively (see Table 7.7).

Authors' activities are expected to remain essentially the same regardless of size of journal. An exception might be if rejections and requests for revisions varied substantially, necessitating more or less follow-up work on the part of an author. In addition, small journals sometimes have relatively smaller support staffs, thus placing a greater responsibility for editing and graphics on the author.

Differences in the publishing process may be substantial among the three sizes of journals. Generally small journals may find it necessary to economize to keep their prices at a reasonable level; they may also need to counteract the lack of any scale economies. The number of pages would be expected to be less, a hypothesis that is confirmed by Leeson

Table 7.7 Circulation data for small, medium, and large journals, 1977.

Journal Size	Individual	Institutional	Foreign	Total
Subscriptions per journal				
Small	437	626	328	1,391
Medium	2,813	1,468	1,173	5,454
Large	17,307	7,069	3,075	27,450
Total number of subscriptions (millions)				
Small	0.99	1.42	0.75	3.16
Medium	4.44	2.32	1.85	8.62
Large	10.31	4.21	1.83	16.36

and Machlup[4] in a special analysis performed by them for us.* Their data also indicate that other economies are utilized by small journals, with an editorial cost per page of $10.55 for them as compared with a per-page cost of $33.51 for medium and large journals. Among other factors, this difference probably reflects more volunteer efforts for small journals.

More than balancing the lower editorial costs, small journals must recover their fixed costs from a much smaller number of subscribers. Thus prices for small journals are considerably higher than those for larger journals. Chapter 3 shows 1977 institutional average prices of $41.62 for small journals, $31.83 for medium journals, and $21.47 for large journals. The difference in prices is likely to be one factor affecting the subscription rates charged individuals and institutions.

Also affecting differences in subscription rates by individuals and institutions is the fact that there are fewer libraries than scientists, and generally fewer libraries interested in a particular journal than there are individuals. This may be viewed as equivalent to an upper limit on institutional subscriptions, regardless of the total number of subscriptions sold, and its results are reflected in the subscription data, which show 45 percent of small journal subscriptions and about 27 percent of medium and large journal subscriptions going to libraries (Table 7.7). The foreign audience for a particular journal is also relatively fixed, and percentages of foreign subscriptions range from 24 percent for small journals to 11 percent for large journals.

Proportionately more small journals go to libraries, necessitating the conduct of various processing activities there. Small journals may be retained longer by libraries, since there are fewer interlibrary loan

*In 1974, the small journals in Leeson and Machlup's sample had an average of 1,068 pages, the medium journals an average of 1,752 pages, and the large journals an average of 1,764 pages.

Table 7.8 Availability and use of all source of copies of journals, by size of journal, 1977.

Journal Size	Articles Written (000)	Uses (millions)	Uses per Article	Article Copies Distributed (millions)	Uses per Article Copy Distributed
Small	147.1	29.5	200	148.3	0.20
Medium	170.4	68.5	402	628.8	0.11
Large	64.4	146.0	2,267	1,315.5	0.11
Total	382.3	244.0	638	2,092.6	0.12

sources available. Because libraries expect them to be used less, small journals may undergo somewhat less processing than the larger journals. It has been hypothesized that both large and small journals are more likely to be the subject of interlibrary loans than the other category.

A&I coverage of large journals is likely to be greater than that for small journals. It is not known how the proportion of articles identified through A&I services varies with size of journal, but some differences could be expected. Generally an A&I service is used to identify relevant articles in journals not regularly scanned, and these journals will be both large and small.

Size of journal does not appear to affect activities associated with individual subscriptions, beyond the differences in the proportion of such subscriptions. With reprints and preprints, it may be hypothesized that more of these article forms are obtained by and requested from authors and from the publisher with large journals than with small.

Our use estimates by size of journal used for cost purposes are based on circulation and thus reflect the hypothesis that journals are used in proportion to the number of available copies. An alternate, or possibly added, hypothesis might be that articles in smaller journals are more likely to be of interest to the small, presumably more homogeneous groups receiving them and thus readership per available copy might be higher. Since overall we have seen that the volume of activity per available copy is somewhat greater for small journals, only a higher readership per copy would make per-use costs comparable.

The availability of copies from small, medium, and large journals is given in Table 7.8. The table also shows the total distribution and use figures for small, medium, and large journals. As expected, large journals are estimated to have considerably more uses per article than medium and small journals. Small journals have about 150 uses per article, large journals over 2,800. In terms of uses per copy distributed, estimates are about eight copies distributed for each use for medium and large journals and five copies distributed for each use for small journals. Our use data by size of journal were developed from subscription data, assuming proportional readership for each type of subscriber, so the differences in per-copy use figures reflect only differences in the breakdown of subscribers for each size of journal.

THE IMPACT OF TIME ON JOURNAL FLOW

Many of the activities associated with journals take place within a relatively short period of time, slightly over one year. Garvey suggests that the actual writing of an article takes about five months.[5] The time from submission of the article to a journal to publication, called publishing delay, averages about ten months. Data on this period, which includes all of the reviewing, revision, editing, and journal production activities, are shown in Table 7.9.

These delays are short relative to the extended period of time—over twenty years or more—over which an article may be read and cited. The distribution of uses by age of article obtained from our user survey is indicated in Table 7.10. The survey was conducted in spring 1978, with about two-thirds of the respondents indicating use of a very recently published article. Another 20 percent of the articles read had been published in the two previous years, and the remaining 16 percent were dated 1975 or earlier, some going back prior to 1960. These data suggest the dual nature of the journal: for use as a current awareness mechanism and as a permanent record of scientific accomplishment. When the first role is considered, the publishing delay previously referred to as relatively short becomes a matter of more concern.

The distribution of citation uses, as a subset of all readings, extends over a longer period of time and reflects a considerably longer half-life. Half-lives observed in various fields of science, for both citation and all uses, were displayed in Table 6.9. Over all fields, the half-life for citations is 10.8 years and the half-life for all uses is 0.2 years.

COMMUNICATION FLOW BETWEEN THE
UNITED STATES AND OTHER COUNTRIES

Throughout the journal system there is a significant level of interdependence between U.S. participants and participants from other countries. In our data collection efforts, our primary concern has been with U.S. journals and U.S. scientists and engineers, but we have also

Table 7.9 Publishing delay, by field of science, 1977 (in months).

Field of Science	Mean Publishing Delay
Physical Sciences	8.0
Mathematics	20.5
Computer Sciences	10.6
Environmental Sciences	14.4
Engineering	9.0
Life Sciences	12.1
Psychology	12.1
Social Sciences	10.3
Other Sciences	5.8
All Fields	10.3

Source: King Research, Inc. Author Survey.

Table 7.10 Distribution of article readings, by year of publication.

Year of Article Publication	Number of Readings (millions)	Proportion of Readings (%)
1978	157.8	65
1977	41.0	17
1976	7.3	3
1975	2.4	1
1970–1974	26.8	11
1960–1969	4.9	2
Prior to 1960	3.9	2
Total	244.0	100

Source: King Research, Inc. User Survey.

been able to determine the magnitude of many of the interactions with foreign journals and foreign scientists and engineers. Data are available on journal authorship, publishing, and use.

In the area of authorship, a simple model suggests four major classes of articles: combinations of both U.S. authors and foreign authors publishing in both U.S. and foreign journals. An article may involve coauthors from different countries, and a single author may move or reside and work in (or may be a citzen of) different countries. A number of journals are published concurrently in more than one country. Accommodating these exceptions, however, we return to the four-cell classification for our analysis.

Earlier estimates indicate that there were about 382,000 articles published in U.S. journals in 1977, with some 295,000 of these authored

Table 7.11 Number of articles written by U.S. and foreign authors, by field of science, 1977.

Field of Science	U.S. Journals		Foreign Journals
	U.S. Authors	Foreign Authors	U.S. Authors
Physical Sciences	33.7	23.8	9.9
Mathematics	5.8	2.7	0.6
Computer Sciences	6.8	2.2	0.6
Environmental Sciences	13.2	3.8	2.3
Engineering	32.3	14.4	6.4
Life Sciences	95.0	32.9	14.8
Psychology	11.8	1.6	0.4
Social Sciences	55.1	2.0	4.7
Other Sciences	41.4	3.8	2.6
All Fields	295.1	87.1	42.2

by U.S. scientists. U.S. authors, in the author survey, also reported publishing about 42,000 articles in foreign journals in 1977. The total number of articles published in foreign journals by foreign authors is not known. We can see however, that about twice as many foreign articles are published in U.S. journals as U.S. articles published in foreign journals. This ratio is somewhat low relative to the numbers of U.S. and non-U.S. scientists and may reflect the greater prestige of U.S. journals and/or U.S. science activities.

Estimates of U.S. and foreign authorship are indicated by field of science in Table 7.11. Psychology and the Social and Other Sciences are quite low in the proportion of U.S. published articles written by foreign authors; they are also low in foreign-published articles written by U.S. scientists, suggesting a relative isolationism in these fields. Foreign Mathematics journals are also low in U.S.-authored contributions, but U.S. journals in the field publish a high proportion of foreign contributions. Other fields with high levels of foreign authorship are the Physical Sciences, Engineering, and the Life Sciences.

Publishers receive articles from authors in different countries and also have subscribers worldwide. Average subscription volume for U.S. journals by field of science is shown in Table 7.12, indicating a 16 percent rate of foreign subscriptions overall with higher percentages in the Physical Sciences and Engineering. U.S. libraries and individuals also subscribe to non-U.S. journals. Fry and White, for example, have found the proportion of foreign periodical holdings in U.S. libraries to range from 2 percent in small public libraries to over 50 percent in large academic libraries.[6]

Many of the A&I publications and data bases are international in character, covering all of the world's literature in their particular subject area. *Chemical Abstracts,* for example, includes coverage of about four times as many foreign publications as U.S. publications. Other major data bases have similarly high proportions of foreign coverage. On the use side, subscriptions to A&I publications and data-base searches are also international in nature.

Data on use of foreign materials have previously been restricted primarily to citations, and several significant studies have been done in this area.[7,8] These include the development of citation indexes that measure the extent to which the literature of a particular country is cited relative to the volume of citation expected based on the country's share of the world's publications. Using this method, an index of 1.0 means that there were as many citations to a country's literature as would be expected from its share of the world's publications. When citation indexes are computed for various scientific fields, the United States consistently ranks first or ties for first place among the major countries of the world. The United States and the United Kingdom consistently have indexes of 1.0 or greater, and the indexes of West Germany, France, and the Soviet Union are 1.0 or greater in at least one field.[9]

The user survey conducted as part of the project reported on in this book yields additional data on journal interchanges, covering all types of use. For U.S. scientists, we have identified both the total volume of use of the journal literature and also the country of publication and country of authorship of the articles used. Over all fields, 7 percent of readings are from foreign journals and 14 percent are of articles written by foreign authors. Variations among the fields of science are indicated in Table 7.13, which gives the number of readings by U.S. scientists of U.S.- and foreign-published articles by U.S. and foreign authors. In each field, the level of reading of foreign-authored articles is higher than that for foreign-published articles. Physical Sciences, Mathematics, and Computer Sciences have higher levels of use of foreign materials, while Psychology, the Social Sciences, and Other Sciences have the lowest levels.

In studying authorship, it was determined that about 23 percent of articles published in U.S. journals were by foreign authors. Table 7.13 indicates that 9 percent of the readings of U.S. journals (by U.S. readers) involved foreign authors. Similarly about 13 percent of articles authored by U.S. scientists are published in foreign journals, but readings of U.S.-authored articles in foreign journals make up only 7 percent of all such readings. This suggests a relative emphasis by U.S. readers on U.S.-authored and/or U.S.-published articles. Consistent with this finding is a parallel one involving citation indexes, in which scientists can be

Table 7.12 Average journal subscription volume by type of subscriber, by field of science, 1977.

Field of Science	Subscriptions per Journal				Proportion Foreign (%)
	U.S. Individuals	U.S. Institutions	Foreign	Total	
Physical Sciences	1,177	1,229	1,321	3,727	35
Mathematics	2,435	1,602	239	4,276	6
Computer Sciences	6,643	7,153	2,923	16,719	17
Environmental Sciences	7,394	3,180	1,503	12,077	12
Engineering	3,477	2,862	2,234	8,573	26
Life Sciences	4,270	1,429	840	6,539	13
Psychology	1,154	1,654	522	3,330	16
Social Sciences	1,202	833	281	2,316	12
Other Sciences	11,830	4,474	2,556	18,860	14
All Fields	4,201	2,126	1,000	6,327	16

Table 7.13 Readings of U.S. and foreign journal articles by U.S. scientists and engineers, by field of science, 1977.

Field of Science	Total No. of Readings Annually (millions)	Readings of U.S. Journals by U.S. Authors (millions)	Readings of U.S. Journals by Foreign Authors (millions)	Readings of Foreign Journals by U.S. Authors (millions)	Readings of Foreign Journals by Foreign Authors (millions)	Proportion of Foreign Journal Readings (%)	Proportion of Foreign-Authored Readings (%)
Physical Sciences	39.3	29.7	5.8		3.8	10	24
Mathematics	5.9	4.4	0.6	0.2	0.7	15	22
Computer Sciences	10.9	8.3	2.3	0.3		3	21
Environmental Sciences	8.2	6.9	0.5	0.1	0.7	10	15
Engineering	82.6	71.2	6.1		5.3	6	14
Life Sciences	49.7	41.1	3.4	1.3	3.9	10	15
Psychology	11.5	10.8	0.5	0.1	0.1	2	5
Social Sciences	28.3	27.6	0.5		0.2	1	2
Other Sciences	7.6	7.4	0.2				3
All Fields	244.0	207.4	19.9	2.0	14.7	7	14

seen to cite work done in their own country with greater relative frequency than work done in other countries. With use, as in other areas, the other side of the picture—foreign use of U.S. publications—is not known.

REFERENCES

1. W. D. Garvey, N. Lin, and C. E. Nelson, "Some Comparisons of Communication Activities in the Physical and Social Sciences," in *Communication Among Scientists and Engineers,* ed. C.E. Nelson and D. K. Pollack (Lexington, Massachusetts: Lexington Books, 1970).
2. N. Lin, W. D. Garvey, and C. E. Nelson, "A Study of the Communication Structure of Science," in *Communication Among Scientists and Engineers.*
3. King Research, *Library Photocopying in the United States: With Implications for the Development of a Copyright Royalty Payment Mechanism* (Washington, D.C.: Government Printing Office, 1977).
4. K. W. Leeson and F. Machlup, "A Catalog of Statistical Data on the Economic Characteristics of Primary Research Journals in the Sciences and Technology," Preliminary Findings, March 18, 1977.
5. Garvey, Lin, and Nelson, "Some Comparisons of Communication."
6. B. M. Fry and H. S. White, *Publishers and Libraries: A Study of Scholarly and Research Journals* (Lexington, Massachusetts: Lexington Books, 1976).
7. J. D. Frame and J. J. Baum, "Cross-National Information Flows in Basic Research: Examples Taken from Physics," *Journal of the American Society for Information Science* 29(5):247–252 (1978).
8. National Science Board, National Science Foundation, *Science Indicators, 1974* (Washington, D.C.: Government Printing Office, 1975).
9. N. K. Roderer and C. G. Schell, *Statistical Indicators of Scientific and Technical Communication (1960–1980): Worldwide Indicators* (Rockville, Maryland: King Research, 1977).

Chapter 8

Economics of the Scientific and Technical Journal System

The previous chapters have described the structure of the scientific and technical journal system from generation to use. The last chapter dealt with the flow of information, the systemic relationships among participants, and the implications of these relations on the journal system. This chapter describes economics of the journal system and the economic relationships among the participants.

AN ECONOMIC FRAMEWORK FOR JOURNAL SYSTEMS ANALYSIS

In Chapter 1 we briefly described the framework on which the analyses in this book are based. Here we elaborate on the framework to demonstrate its economic implications for the journal system. This framework is based on subdividing the journal system into functions, products and services that perform the functions, and all of the activities necessary to produce products or operate services. Each activity is described in terms of input, process, and output variables. Input variables include resources such as quantities of materials and personnel. The process variables include types of equipment and level of personnel expertise. The output variables include such factors as number of items produced during the activity. The difference between input and output is the value added by the activity.

Activities can also be defined in terms of their performance and cost. This relationship between process variables, cost elements, and performance for each activity is characterized by the schema in Figure 8.1. Process variables include the type of equipment, personnel, and supplies that are used in the processes necessary to perform each activity. The equipment can be characterized by its general type (for example, hot lead typesetting versus computer-controlled photocomposition) or by specific brand or model. Personnel can be described by their experience, training, or expertise. In each instance,

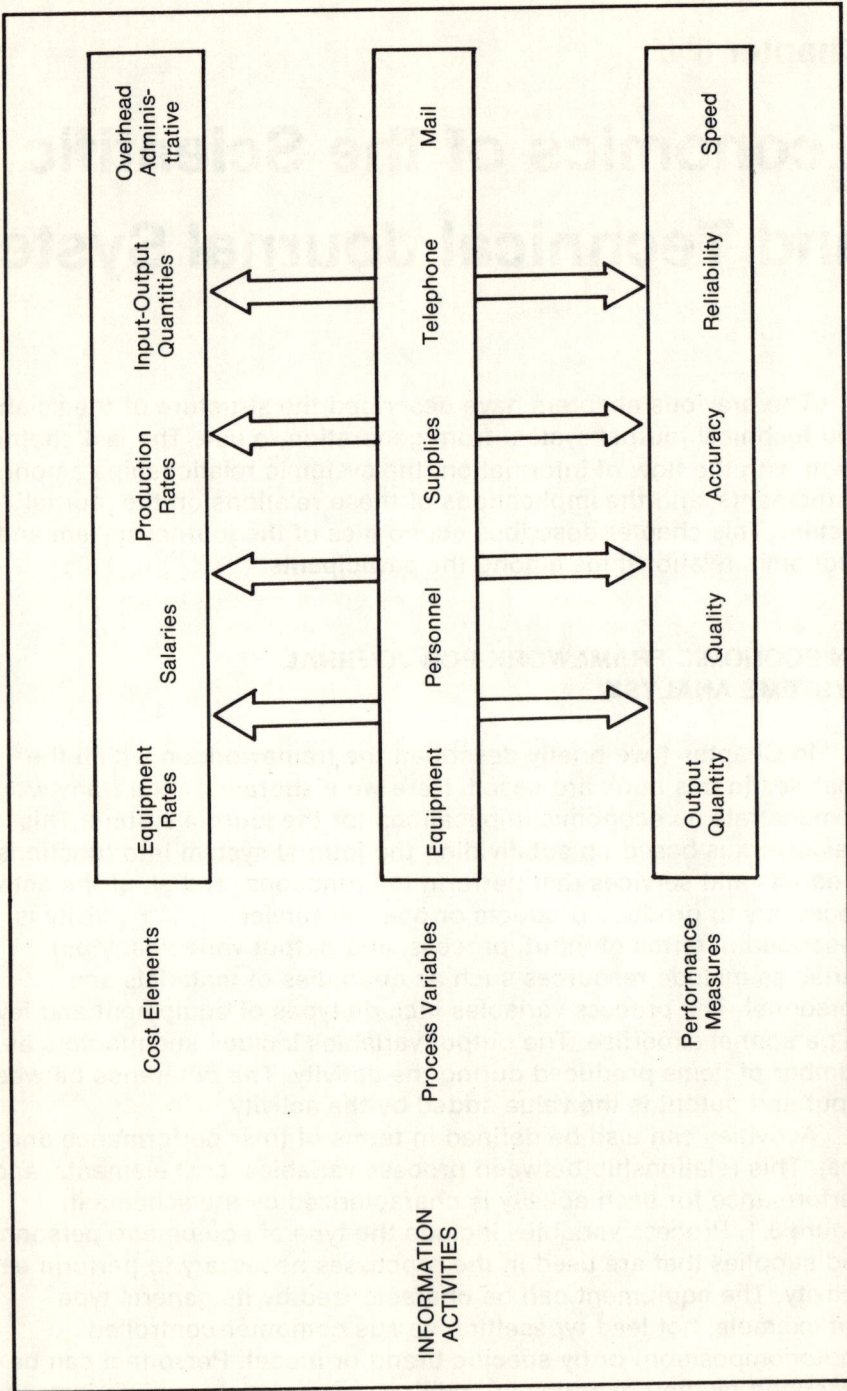

Figure 8.1 Process variables, cost elements, and performance measures associated with information activities.

the equipment, personnel, and supplies can be characterized along dimensions related to cost or performance.

Cost elements are those items that determine the expenditures for an activity. They include such measures as equipment rates and capacity; personnel salaries, rate of production, and capacity; unit cost of supplies; and the number of units of input and output. Cost elements also include allocations of overhead, administrative, and facility costs to an activity. The cost elements can be formulated in a model that will yield total cost under a variety of process variables or with changes in values of the cost elements. An activity can also be described in terms of its performance, such as in terms of quality, speed, reliability, accuracy or output quantity.

Both the cost elements and the performance measures are affected by changes in the activity process variables, so they are under the direct control of management. Almost any change in input, process, or output will have a direct effect on cost and performance. The relationship of an activity's cost and performance is often referred to as its efficiency or productivity. Efficiency is often expressed in terms of output quantities per dollar. Productivity, on the other hand, is often given in terms of output quantity per unit of a cost element. Productivity, for example, could be output quantities per time unit of labor, such as number of abstracts written per hour of labor or keystrokes per hour of labor and/or equipment.

On a broader scale is an entire information product or service, which is usually the sum of several activities. Here we find three general types of measures (Figure 8.2). Description of an information product or service involves its costs, its attributes, and the effectiveness of all the activities that go into it. The cost is the sum of the costs of all activities. Each information product and service can be described by several attributes, such as the performance measures, price, and specific characteristics. The effectiveness measures are those related to the users of the product and service, such as number of uses, amount of repeated use, and user satisfaction. Effectiveness measures should be directly related to activity performance measures and other attributes, such as price or journal characteristics. That is, measures of user satisfaction will depend partially on performance factors, such as article quality or speed of publication, or general characteristics, such as style of publication or number of issues per year. The number of uses is also related to such factors as user satisfaction, price, ease of accessibility, or awareness. In modeling, a common approach is to use multiattribute analysis where an effectiveness measure is the dependent variable and performance measures and the other factors mentioned above are the independent variables.

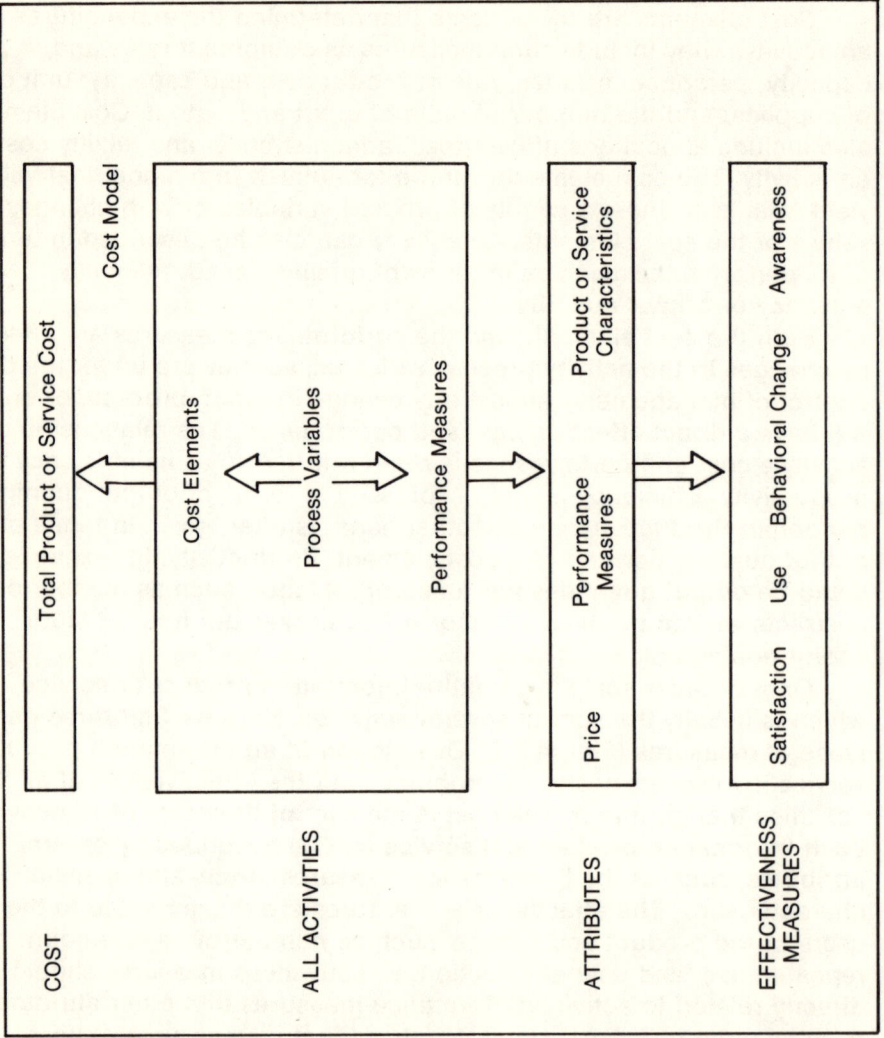

Figure 8.2 All activities, cost, attributes and effectiveness measures associated with information products and services.

It is common to describe information products or services in terms of cost effectiveness. In this chapter we compute cost per use of journals by participant contributions and different paths with different levels of effectiveness.

Another economic comparison of cost and effectiveness involves the relationships of price and demand versus the cost and demand. This comparison is the basis for traditional economic analysis of marketed goods involving buyers and sellers. In the journal system, it serves as a basis for analysis of the economic interface between two system participant groups such as publishers and libraries or publishers and individual subscribers.

Although effectiveness measures depend to some degree on performance, they are not under the direct control of management since these measures are made largely from the perspective of users and management cannot control users or their behavior. Similarly the total cost is not entirely under the direct control of management since part of the cost is a function of the number of uses.

Broadening the perspective even further, we describe a specific system function. Measures describing functions are shown in Figure 8.3. A system function may be performed through several journal products or through several participants. For example, an article could be distributed in different forms by several participants such as by a personal journal subscription distributed by the publisher, a photocopy distributed by a library, or a reprint distributed by the author. The economic demands for these alternatives are interdependent. If the price, quality, or availability of one alternative is unsatisfactory, a substitute product or service may be made. Thus changes in attributes of a journal product (or service) will have some effect on all of the other products (or services) that are used to perform a system function.

Some subtle interactive effects can occur in such a situation.[1] For example, a difficult economic situation exists with respect to journals, particularly small journals with a low number of subscriptions. Libraries are dropping some of them from their subscription lists because they do not feel that they can afford journals that have low use. The break-even point on purchasing a journal as opposed to interlibrary loan appears to be about five to fifteen uses per issue. If a journal is used fewer than that many times in a year, it is less costly for the library to borrow a copy each time than to purchase a subscription to the journal.

The problem is that there is a large fixed input cost associated with publishing journals, a cost that must be partially absorbed by library subscription prices. If all libraries canceled their subscriptions to small journals when usage is under the break-even point in the library, the price of these journals would have to increase substantially in order

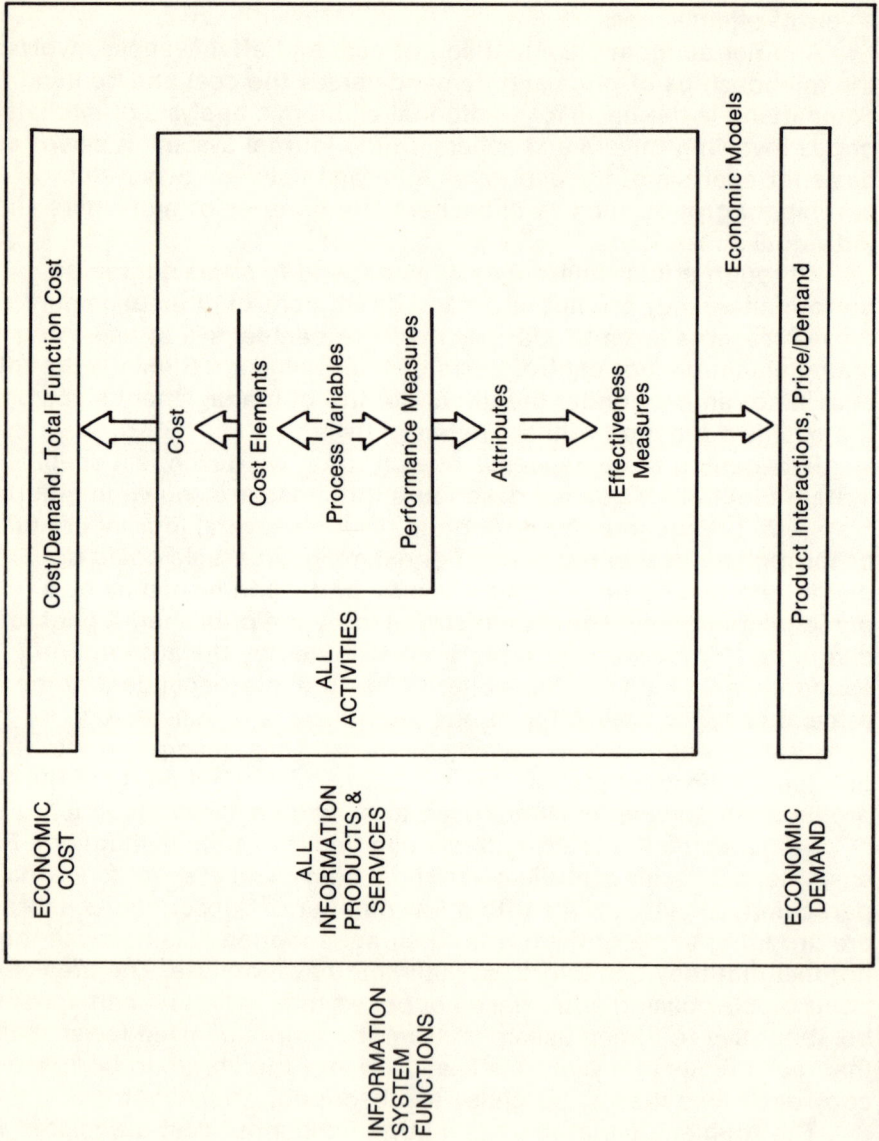

Figure 8.3 All information products and services, economic cost, and economic demand associated with information system functions.

to recover the large input cost. Thus the libraries that have a sufficient number of uses would pay more, and the remaining libraries would obtain copies through interlibrary loan. As the price increases, the break-even point does as well, so more subscriptions could be canceled. There is a strong possibility that some small journals might not survive this snowballing effect. Yet there are many important articles that have a small audience and should be made available in the same manner.

One of the important aspects of the journal system is the amount of organizational interdependence among participants. This can present a substantial problem because the motives and incentives of these participants are often at cross-purposes. An author may submit an article because of the need to publish in some environments. Yet possible low quality of some such articles would conflict with the needs of other participants. Commercial publishers would like to maximize their profits. However, increasing price will reduce the purchase and use of journals, possibly resulting in redundant research or other negative consequences. Thus one of the most difficult aspects of journal systems to study and to describe is the interaction among system participants. It is, accordingly, the least understood.

Some of the greatest potential improvements in the journal system could come from understanding and optimizing the interaction among system participants. At the very best, an appreciation of the economic and systemic interdependencies among participants should result in a better overall system. Perhaps, with such an appreciation, the antagonisms of librarians and publishers would not have occurred in connection with the new copyright law.

The broadest level of consideration for complete description is the entire journal system (Figure 8.4). In journal publishing, this is the entire system represented by the spiral of functions depicted in the first chapter. It involves the participants shown in Figure 1.4, as well as other influencing groups, such as government policy makers and regulatory agencies. The total costs of a system include all direct and indirect costs in the system. For example, in a comparison of electronic alternatives to paper-based publishing of scientific and technical articles, we have included author preparation costs, reviewer costs, and reading costs, all of which contribute considerably to the total costs. The costs, regardless of whether they are direct, hidden, or donated, involve all participants. The analysis of total costs given in this chapter begins with the specification of cost elements for every identifiable activity in the system. The cost elements are combined by means of models to form information product or service costs, which are summed to yield total costs.

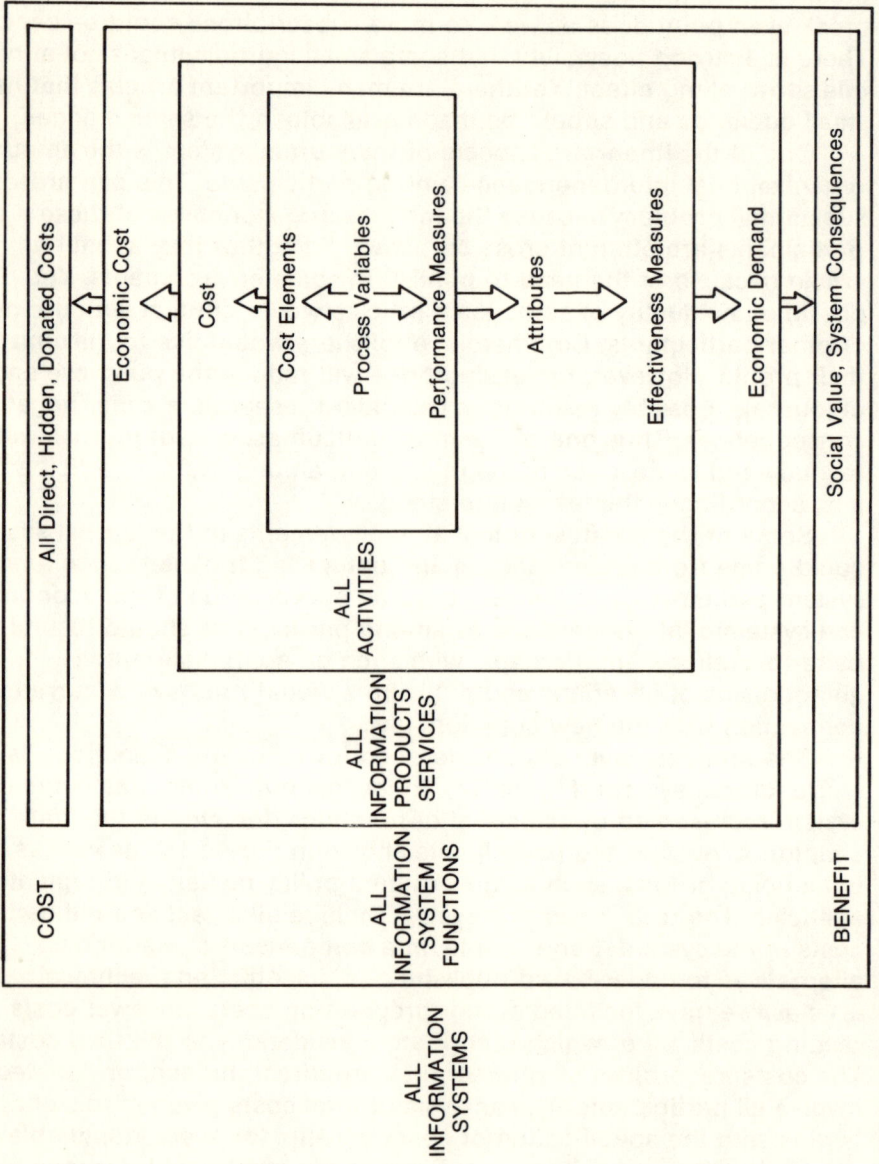

Figure 8.4 All information system functions, costs and benefits associated with an information system.

We consider benefits to be the consequences of the system. In a scientific and technical journal system, we feel that these consequences should be considered in terms of their contribution to society (for example, facilitating advancements in medicine). However, such social benefit is probably impossible to measure.[2] Faced with the difficulty, we maintain that one should attempt to establish measures that are at least correlated to benefit. We consider these are surrogate measures. Just as some performance measures are surrogate measures for effectiveness (use), use is a surrogate measure for benefit. Presumably more extensive use implies greater benefit, although this depends somewhat on the purpose of the use. Reading articles for general knowledge would seem to have less value than for direct application to research. The point to emphasize is that any measure of performance, effectiveness, or economic demand should be chosen based partially on its apparent relationship to benefit, including qualitative statements of the journal system's mission, goals, and objectives. Since we cannot give measures of both cost and benefit in this book, we have relied on cost per use.

THE COSTS OF SCIENTIFIC AND TECHNICAL JOURNALS

The costs associated with journal communications can be identified from a number of different perspectives or dissaggregated in several different ways. For example, one might want to consider the function performed, the source of funds, or the type of cost (such as labor or equipment) as the basis for our analysis and present costs according to breakdown of these elements. The costs are, of course, collected mainly in terms of the contributions of various system participants, and this would also be a way to present them. To permit the widest range of analyses, costing procedures are based on an extensive list of activities performed by the system participants. These activities, discussed in Chapters 3 through 7, cover all major activities concerned and were designed to serve in a general model of the current system, which could be modified to reflect a wide range of alternatives or trends over time.

While the cost for the various activities can be aggregated in many ways, perhaps the most useful presentation is to consider the four major participants involved. These participants—scientists as authors and users, publishers, libraries, and secondary organizations—can be shown to encompass all activities performed. Tables in this section present costs according to these categories. For some purposes, it is necessary to have finer breakdowns of cost, as shown in Table 8.1. This

Table 8.1 Activity cost data.

Participant	Aggregation in This Section	Aggregation in Previous Chapters
Authors	Authorship	Writing, typing, dispatching manuscript
		Acquisition of reprints for distribution to fellow researchers
Publishers	Prerun	Editing (staff)
		Editing (donated)
		Review
		Typesetting
	Runoff	Runoff (reprints and preprints)
		Runoff (subscribers)
	Library	New journal acquisitions
Libraries and secondary organizations		Maintenance
		Storage
		Library use (circulation, in-library, ILL)
	A&I	A&I production
	Preuse	Acquisition (subscription, preprints and reprints)
		Storage (subscription, preprints, and reprints)
		Individual search
		Data-base search
Users		Photocopying
	Reading	Reading

breakdown is the minimum set necessary to consider different participants, paths, and fields of science in a general sense. The category associated with authors includes all activities related to authorship, activities of publishers are divided into prerun and runoff publication costs, libraries and abstracting and indexing service activities are broken out separately, and the user category is shown as including preuse activities and actual reading.

Before presenting the actual costs,* an important consideration in the analysis of journal communication costs is the matter of timing. Use of a particular journal article may extend over a number of years. On the average, measurable use will continue over at least fifteen years. If we are considering, say, 1960 journals, we must monitor their use through 1977 and possibly beyond. Costs associated with the journal will include a number of initial items (including author-related activities, publishing, library activities, abstracting and indexing, and user activities) and the ongoing storage and use of the journal over its lifetime. To simplify matters, we can treat the initial costs as occurring in the year of publication. Use costs, however, will increase substantially over time. Our estimate for preuse activities and the time spent on reading a journal article in 1977 is $11.90, while in 1985 the cost would be $14.40, in dollars of 1977 purchasing power. The 21 percent increase reflects system changes and accelerated growth in professional sources. For such reasons we must carefully identify the year in which a given cost will be incurred.

For some purposes it is important to know all communication costs associated with a particular set of journal articles—for example, those published in a given year. From these estimates, total cost per use† can be derived, and the effect of changes in processing methods can be observed. What cannot be readily determined by this method, however, is the expenditures that will take place each year. These costs, observed over the years, give a more realistic picture of the results of modifications of the system. The method employed for most of the economic considerations that follow, then, is to identify all costs incurred in any particular year. Then these costs can be broken down according to the publication year of the journals covered.

Cost per use figures are also presented here. These data must be interpreted with caution, since they represent all costs incurred in a given year distributed over all uses in that year. Thus, if a 1977 journal is used in 1980, the cost associated with that use will include actual use costs and an allocation for the production of 1980 rather than 1977

*It is emphasized that the costs given in this book are not actual costs and, as such, merely represent an order of magnitude of what the true costs may be. Even so, we believe that cost comparisons are useful to understand the journal system.
†Numbers of uses are estimated from a user survey described in Chapter 6.

journals. This is acceptable because a high proportion of use occurs in the first year and, for the remainder of uses, the system is very stable. If, however, substantial changes are to be considered, other modeling approaches should be utilized to determine per use costs.

Another consideration underlying our economic analysis is the allocation of costs associated with scientific and technical communication outside the United States, for there is a significant amount of interchange among various countries throughout the communication cycle.[3] For example, over 20 percent of the journal articles published in the United States are written by non-U.S. authors (and, in turn, U.S. authors are known to publish in non-U.S. journals). Materials published in the United States are used by scientists of all countries, and U.S. scientists use materials from other countries as well. Thus it is difficult to identify precisely the boundaries of U.S. communication activities. For our purposes, we have chosen a very restrictive definition; we include only costs incurred in the United States and associated with U.S. scientists' use of U.S. materials. Excluded are costs associated with foreign authorship for U.S. journals and all use by U.S. scientists of foreign published materials.

Total Journal Communication Costs

The communication costs associated with all 1977 scientific and technical journals are shown in Table 8.2. The costs are categorized under the major participants: authors, publishers, libraries and secondary organizations, and users.

Author costs represent all those incurred by authors and their support staff in writing and submitting a manuscript to a publisher, revising it when required, and proofing the typeset article. These vary among journals according to the salary of the author and the support staff persons, by the magnitude of revisions required, and by the number of proofs sent to the author. On the average, the cost of authorship for a 1977 article was just under $1,800, with a total for all U.S.-authored articles (78 percent of all articles) of $534 million.

Publishing costs cover the publication and initial distribution of subscription copies and reprints. Included are prerun, or first-copy costs, and runoff costs. Prerun costs include refereeing, editing, and composition costs incurred by the publisher or by volunteer editors and/or referees. Since the publishing costs are incurred in the United States, they are included in our calculations in their entirety. An alternate method would be to allocate the costs to the United States and other countries based on the number of subscriptions distributed. The difference is relatively minor in the overall calculations; total prerun

Table 8.2   Total costs of communication through scientific and
technical journals and costs per use, by participants,
1977.

Participant	Total Cost (millions)	Cost per Use (244 million uses)
Authors		
Authorship	$534	$2.20
Publishers		
Prerun	397	1.60
Runoff	221	.90
Subtotal	618	2.50
Libraries and secondary organizations		
Library	434	1.80
A&I services	45	.20
Subtotal	479	2.00
Users		
Preuse	551	2.30
Reading	2,354	9.60
Subtotal	2,905	11.90
Total	$4,535	$18.60

costs are estimated at about $1,050 per article in 1977, and this figure
would be $890 with the foreign allocation excluded. The total for all
journal articles is $397 million.

Runoff costs include printing and distribution by the publisher for
subscription copies and reprints. These costs come to $0.09 and $0.35
per article copy distributed for the two media, respectively. Costs for
both U.S. and foreign subscriptions copies are included in the
calculations, as are all reprint costs. Total runoff costs included in the
model are about $49,700 per journal, or $221 million.

Libraries acquire and process journals, maintain and store them
over long periods of time, make them available for use, and provide user
support services. These operations are costly. New acquisitions, on the
average, cost nearly $100 (excluding subscription price), and annual
maintenance of a journal is estimated at about $30. An internal use in
the library accounts for $2 and obtaining a copy of a journal article from
another library (interlibrary loan) costs $20 when both the borrowing
and lending libraries' activities are considered.[4] Together library costs
incurred in 1977 came to $434 million, with $290 million of this
representing processing and storage costs and the remaining
$144 million representing use costs.

Abstracting and indexing services combine descriptions of
published journal articles into indexes, which serve as one means of
identifying journal literature of interest. The estimated cost of U.S.

abstracting and indexing of the U.S. S&T journal literature in 1977 was $45 million (based on an estimated 3.2 million items processed). Together with library activities, this yields a total of $479 million.

The fourth category is use, which accounts for more than 60 percent of the total communication costs. A major portion of this is made up of actual reading costs, calculated at $9.60 per reading in 1977, for a total of $2,354 million. Other costs included in the use function are those associated with the individual's search for materials ($205 million), with machine-readable data-base searches ($107 million), and with photocopying ($47 million). In the case of individuals, steps taken to provide access to a journal include taking out a subscription and/or requesting reprints and preprints and storing copies received. Acquisition is the more costly of these operations, and allowing $2 (fifteen secretarial minutes) for each individual subscription or request yields a total cost of $139 million. Storage of all copies of journals acquired over a fifteen-year period amounts to $52 million. Total use costs are $2.9 billion, or $11.90 per use.

The communication function of transmission is not included explicitly in these figures, but transmission activities (such as mailing) are covered by each of the four participant groups. Authorship, for example, includes the mailing of an author's manuscript to the publisher, possible mailings to other publishers if the article is rejected, and mailings of changes and proofs back to the publisher when the article is in the production cycle. Publication involves transmissions from the publisher to other system participants—authors, referees, possibly the printer, and, of course, subscribers. Libraries cover many transmission activities, including movement of a journal issue within the library, to and from other libraries, and to the user. Secondary organizations are both users of journals, receiving them from the publisher, and publishers of secondary journals, which are in turn transmitted to libraries and individuals. Use, particularly through a library or colleague, also involves transmission activities. Most of these activities are closely tied to the other function that they support, and it is difficult to make any useful distinctions in the costs. The total communication costs by participant are given in Table 8.3.

Summing up all costs associated with journal communications, the total for 1977 is $4.5 billion. This amounts to a total average expenditure of $18.60 for each use made of a U.S. scientific and technical journal article in 1977. By participant, the per-use figures are $2.20 for authors, $2.50 for publishers, $1.90 for libraries and secondary organizations, and $11.90 for users. The costs associated with the user* dominate heavily,

*Author, user, and donated costs (referee or editor) are often paid from funds allocated to research and development.

Table 8.3   Total scientific and technical journal communication costs, by participant, 1977.

Participant	Total Cost (millions)	Proportion
Authors	$534	12%
Publishers	618	14
Libraries and secondary organizations	479	10
Users	2,905	64
Total	4,535	100

accounting for over 60 percent of the total. These costs include those incurred directly by scientists or their support staff (individual acquisition and storage, searching, photocopying, and reading). Reading alone represents over one-half of the total journal communication cost.

The second largest category of costs is that of the publishers as producer and distributor of journals. These costs are estimated at about $618 million, or about 14 percent of the total communication expenditures. Author costs are about $534 million, 12 percent of the total. Library-related costs, including acquisition, maintenance, storage, and service costs in the library and also costs associated with A&I products and services, are about 10 percent ($479 million) of the total.

From the breakdown of costs by participants, we can roughly analyze the effects of a change in one system component on the total system. For example, if publishers were to change their production methods but retain their present distribution format, this modification would involve only the publisher, and thus less than 14 percent of the total costs. Changing internal library procedures might affect some portion of libraries' 10 percent of the total costs. The source of the largest proportion of costs, the time devoted to reading, is also one of the least flexible in terms of altering the process to reduce costs. On the surface, then, the largest potential effect on costs would seem to come from changes that affected not just one participant but allowed others to reduce their costs also. The earlier in the communication cycle such a change is introduced, the greater the potential effect on costs.

Journal Communication Costs by Communication Path

Another interesting breakdown of total communication costs is by the communication path taken—that is, from publisher, library, or author to reader. Here we would expect per-use costs to vary considerably because of the differing activities associated with each path.

Table 8.4   Total scientific and technical journal costs, number of uses, and cost per use, by communication path, 1977.

Communication Path	Total Cost (millions)	Number of Uses (millions)	Cost per Use
Publisher	$2,875	178.6	$16.10
Library	1,046	36.1	33.10
Author	572	29.5	19.40
All paths	4,535	244.2	18.60

For the most part, path costs were derived by allocating total costs appropriately. Authorship, prerun, search, and reading costs were found by assuming that the cost per use was the same for each of the paths. Runoff and preservation activities could be directly associated with the path taken, while A&I service and photocopying costs also varied by path because of variations in the relative volume of use for each path. An example is that 25 percent of articles read in institutional subscription copies are photocopied, while the percentage for uses of individual subscription copies is only 9.

The cost breakdown by communication path is shown in Table 8.4. As anticipated, the per-use totals vary from $16.10 for individual subscriptions, to $19.40 for reprints and preprints, and to $33.10 for libraries and other institutional subscriptions. These are associated with total costs of $2,875, $572, and $1,046 million, respectively. Perhaps the most surprising of these figures is that for reprints (author to user path), which indicates that about 13 percent of all journal communication expenditures are attributable to reprints and preprints. This is a significant percentage for an article form that is often considered to be secondary and informal. Library per-use costs are quite high, probably reflecting the library's role, in at least some cases, as an archive and a resource of last resort. Individual subscription copies (publisher path) are the source of 63 percent of journal communication costs, with institutional subscriptions (library path) accounting for the remaining 23 percent.

These considerations relate to the cost aspects of scientific and technical communication only and do not tie in any consideration of performance, effectiveness, or benefits. In addition, it seems reasonable to expect that no alternative system could reduce costs by more than a relatively small percentage of the total. Different alternatives will also affect the total system costs by greater or lesser amounts, not necessarily in proportion to their current share of costs. Thus it may be that, for example, a relatively minor change in publishing processes may have as great an impact as a modification affecting both libraries and users. Here again, many additional factors require consideration.

Journal Communication Costs by Field of Science

No journal is average, despite our recording of publication costs in that manner. They vary greatly by number of pages, number of issues, type of material, proportion of graphics, number of subscribers, and other characteristics. The use of the communication system also takes on different configurations depending on both the specific journal and the user involved. One way of grouping journals and users to provide somewhat more information is by field of science. Journal communication costs for the nine fields of science are presented in Table 8.5.

The total magnitude of expenditures in the fields of science varies considerably, from $153 million in Computer Sciences to $1,088 million in Engineering. The number of uses varies also, so that per-use costs are as low as $13.20 in Engineering and as high as $54.20 in Mathematics. Generally in the Physical Sciences, Computer Sciences, Engineering, and Psychology costs are low, and those in Environmental Sciences, Life Sciences, and Social Sciences are mid-range. The variations are the result of many complex interactions involving numbers of journals, subscribers, readers, and other factors. The major reason for the considerably higher costs in Mathematics is the reading cost; $43 per use as compared with an overall average of about $10. Excluding reading costs, per-use system costs for the fields range from only $4 in Engineering to nearly $17 in Environmental Sciences. Mathematics journals are now mid-range, at about $12.

Journal Communication Costs by Size of Journal

Marked differences can be observed in the per-unit costs of author, publisher, and library activities related to small, medium, and large journals. These costs vary primarily because of the differing volumes of subscriptions and use. Total and per-unit costs are shown in Table 8.6.

The authorship and prerun portions of the total costs were derived mainly on an article basis. Variations in use per article account for a range of per-use authorship costs from $0.60 for large journals to $6.90 for small journals, the largest percentage difference estimated among

Table 8.5 Total scientific and technical journal costs, number
of uses, and cost per use, by field of science, 1977.

Field of Science	Total Cost (millions)	No. of Uses (millions)	Cost per Use
Physical Sciences	$536	39.3	$13.60
Mathematics	320	5.9	54.20
Computer Sciences	153	10.9	14.00
Environmental Sciences	195	8.2	23.80
Engineering	1,088	82.6	13.20
Life Sciences	961	49.7	19.30
Psychology	165	11.5	14.40
Social Sciences	780	28.3	27.60
Other Sciences	346	7.6	45.50
All Fields	4,535	244.2	18.60

the various functions. Per-use prerun costs range similarly. If finer size breakdowns were utilized, the ranges would be even greater.

Subscription runoff costs include fixed (mostly setup costs) and variable costs based on the number of subscriptions. Thus because of the fixed costs, unit costs for small journals will be somewhat higher than for large journals. If articles from large journals are also read less frequently, however, per-use costs for large journals will also be larger on this account. In the case of preprint runoff costs, on the other hand, our model assumes that the per-use runoff cost will remain constant regardless of the size of the journal. All of these factors combine to produce per-use publisher estimates of $6.50 for small journals, $3.60 for medium journals, and $1.30 for large journals.

Library and individual storage and acquisition costs (the latter included in user costs) depend on the mix of individual and library subscribers for each type of journal. Individuals account for 41 percent of U.S. subscribers to small journals, 65 percent of subscribers to medium journals, and 71 percent of subscribers to large journals. The variation among small, medium, and large journals is not great; library costs are $2.50 for small journals and $1.80 for large journals.

Together the author, publisher, and library costs per use for the three sizes of journals come to $3.70 for large journals, $9.10 for medium journals, and $16.00 for small journals. Use costs remain relatively constant, varying with the proportion of individual subscriptions but not in the cost of actual readership. For large journals, use costs make up over 75 percent of total costs; for small journals, they are less than 50 percent. This difference highlights the need for separate consideration of the three sizes of journals.

Table 8.6 Total scientific and technical journal cost and cost per use, by size of journal, by participant, 1977.

Participant	Small Total (millions)	Small Cost per Use	Medium Total (millions)	Medium Cost per Use	Large Total (millions)	Large Cost per Use
Authors	$205	$6.90	$238	$3.50	$90	$0.60
Publishers	192	6.50	249	3.60	183	1.30
Libraries and secondary organizations	75	2.50	137	2.00	267	1.80
Users	376	12.70	834	12.20	1,695	11.60
All participants	848	28.70	1,457	21.30	2,235	15.30

Trends in Journal Communication Costs

It is of interest to project what the situation will be in the future if no significant changes are affected over the years. In order to estimate the costs of what is essentially the current system for 1980 and 1985, it is assumed that the volume of activity and related costs of the various system components will follow trends established over the last ten to fifteen years. For example, the key variables of number of articles and number of journals published were established on the basis of projections of actual data from the 1960–1977 period. Another key variable, the number of uses, was determined from projections of the number of scientists by assuming that the number of articles read per scientist per year will remain essentially constant.

Considering estimates of the numbers of articles written and the numbers of articles read in 1975, 1977, 1980, and 1985 (see Table 8.7), it seems probable that the number of articles written will grow at a greater rate than the number of uses. Since not only the number of articles written but also the number of copies distributed is projected to increase, the number of available copies will grow at a still faster rate. Thus the number of uses per article written will decrease over the years, and the fixed production costs will be allocated over a smaller number of uses. Under this assumption, it is clear that per-use costs will go up faster than anticipated from the combined rate of inflation for processing one unit of material through the system. This factor, combined with considerable rates of inflation for the key cost parameters, will yield large increases in the per-use costs over the 1975–1985 period. Costs are presented below in both current and constant dollars since inflation should not be confused with real cost.

Major cost variables for the total communication system include scientists' salaries, library costs, and publication costs. Scientists' salaries are of particular importance, since they dictate most of the authorship and user costs, which make up more than 75 percent of the total costs. Projections of scientists' salaries (for all fields of science combined) are shown in Table 8.8.

Scientists' salaries climbed steeply in the early 1960s and fell back somewhat in the early 1970s. Projections, however, suggest a 1975–1980 increase of 37 percent, or 6 percent in constant dollar terms. The 1980–1985 constant dollar increase is expected to be an even greater 16 percent. This factor will have a considerable impact on total system costs.

In publication cost trends, labor costs are a dominant factor. Publication trends also affect library costs, since about one-third of library budgets are spent on materials. The remaining portions of library

Table 8.7 Journal articles written, copies distributed, and uses, 1975, 1977, 1980, 1985.

	1975	1977	1980	1985	Ten-Year Change
Articles written (thousands)	353.7	382.3	416.7	476.8	35%
Copies distributed per article	5,290	5,470	5,530	5,690	8
Total copies distributed (millions)	1,871	2,091	2,304	2,713	45
Number of article uses (millions)	231.4	244	262.7	294.0	27

Table 8.8 Scientists' salaries, 1960–1985.

	1960	1965	1970	1975	1977	1980	1985
Current dollars							
Median hourly salary[a]	$3.77	$4.57	$6.05	$8.33	$9.60	$11.44	$14.52
Five-year increase		21%	32%	38%		37%	27%
Constant dollars							
Median hourly salary[a]	$5.49	$6.15	$6.62	$6.55	$6.86	$6.93	$8.07
Five-year increase		12%	8%	−1%		6%	16%

[a]Based on a forty-eight-hour work week.

Table 8.9 Librarians' salaries, 1960–1985.

	1960	1965	1970	1975	1977	1980	1985
Current dollars							
Salary index[a]	56	72	94	113	125	146	183
Five-year increase		29%	31%	20%		29%	25%
Constant dollars							
Salary index	82	97	103	89	89	89	102
Five-year increase		18%	6%	−14%		0%	15%

[a]1972 = 100.

costs are primarily labor related. An index of librarians' salaries is given in Table 8.9. These increases are smaller than those of scientists' salaries, but a 15 percent constant dollar increase is estimated for 1980–1985.

Other variables also influence trends in the total communication

Table 8.10 Total scientific and technical journal cost and cost per use, by participant, 1975, 1977, 1980, 1985.

Participant	1975 Total (millions)	1975 Cost per Use	1977 Total (millions)	1977 Cost per Use	1980 Total (millions)	1980 Cost per Use	1985 Total (millions)	1985 Cost per Use	Ten-Year Change Total	Ten-Year Change Cost per Use
Current dollars										
Authors	$405	$1.80	$534	$2.20	$696	$2.60	$1,017	$3.50	151%	94%
Publishers	483	2.10	618	2.50	801	3.00	1,054	3.60	118	71
Libraries and secondary organizations	422	1.80	479	2.00	602	2.30	857	2.90	103	61
Users	2,360	10.20	2,905	11.90	3,781	14.40	5,442	18.50	130	81
All participants	3,670	15.90	4,535	18.60	5,880	22.40	8,369	28.50	128	79
Constant dollars										
Authors	318	1.40	382	1.60	429	1.60	565	1.90	78	36
Publishers	380	1.60	442	1.80	493	1.90	586	2.00	54	25
Libraries and secondary organizations	332	1.40	342	1.40	371	1.40	476	1.60	43	14
Users	1,855	8.00	2,076	8.50	2,328	8.90	3,023	10.30	63	29
All participants	2,884	12.50	3,241	13.30	3,620	13.80	4,649	15.80	61	26

Note: Total uses are estimated at 231, 244, 263, and 294 million for 1975, 1977, 1980, and 1985, respectively.

costs, and these were incorporated in the cost model. They include such items as photocopying costs, the number of foreign subscriptions to U.S. journals, postage costs, secretarial costs, and storage costs. One area that does not account for a large proportion of the total costs but does show a marked increase is data-base searching (included in preuse costs). Here, the estimates are for $71 million in 1975, $200 million in 1980, and $334 million in 1985. These increases are largely due to an increase in the number of searches conducted. Cost figures for the total journal communication system derived for 1975, 1977, 1980, and 1985 are shown in Table 8.10. Total costs and costs per use are given for each of the functional areas, as are the corresponding 1975–1985 increases.

ECONOMIC INTERFACE BETWEEN PUBLISHERS AND USERS AND PUBLISHERS AND LIBRARIES

Relationship of Price and Number of Subscriptions

Most scientific and technical journals are sold through subscriptions to individuals or libraries. If the price of a journal is increased or decreased, the number of subscriptions will decrease or increase, respectively. Such a demand relationship is shown in Figure 8.5 for individual subscribers; a linear demand curve is drawn to convey the fact that more persons subscribe to journals with lower subscription prices. However, there is a limit to the number of subscribers even if the journals are free. This is denoted on the figure as maximum quantity demanded, D_M. Also, there is some maximum price above which no one will subscribe to the journal. This price is given as P_M. A similar curve is drawn for library subscriptions in Figure 8.6. In both instances (individual and library subscriptions) the curves are loosely derived from analysis performed by Berg. The actual shape of the curve is very difficult to establish, although attempts have been made to show this relationship between price and number of subscriptions.[5,6] The most common approach to measuring the curves is to observe the price and subscription relationship over a large number of journals in a given year or to observe changes in price (in constant dollars) over time and relate these to number of subscribers.

There are several problems with viewing the journal price and subscription relationship in a simplistic manner. First, information from journals can be obtained in several ways, including through subscriptions by individuals, library copies, reprints, and copies from colleagues. If a subscription price is too high for one way, alternative sources can be employed by readers to obtain the information

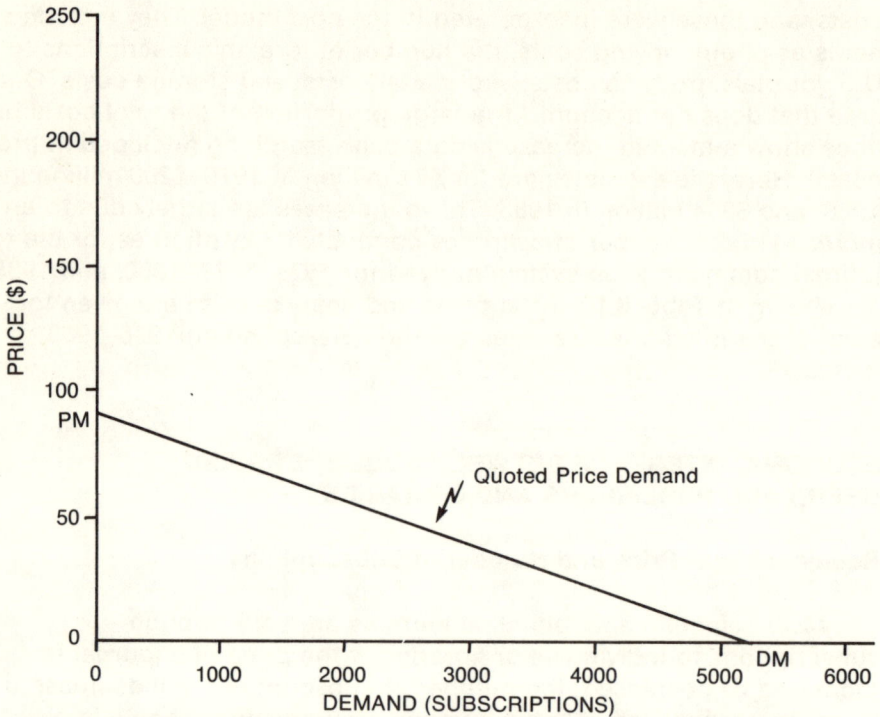

Figure 8.5 Price and demand (subscriptions) relationship for individual subscriptions.

contained in journals. This is a classic example of substitution of equivalent goods.

Second, the price paid for journals often includes other services or advantages such as those found in professional society memberships. As Baumol and Ordover point out,[7] this characteristic, referred to as bundling, makes it difficult to sort out the value attributable to the journal as opposed to the prestige, certification, or social value associated with membership in a professional society.

Third, the perceived quality or value of each journal differs, and the number of potential readers (or subscribers) of different journals varies substantially. One valid way to measure the relationship between price and number of subscriptions is to conduct controlled experiments in which different prices of a journal are quoted to different segments of the population and the number of subscribers is observed at each price level. One such experiment has been conducted with government-sponsored technical reports,[8] but to our knowledge none has been performed with journals or any other information product or service.

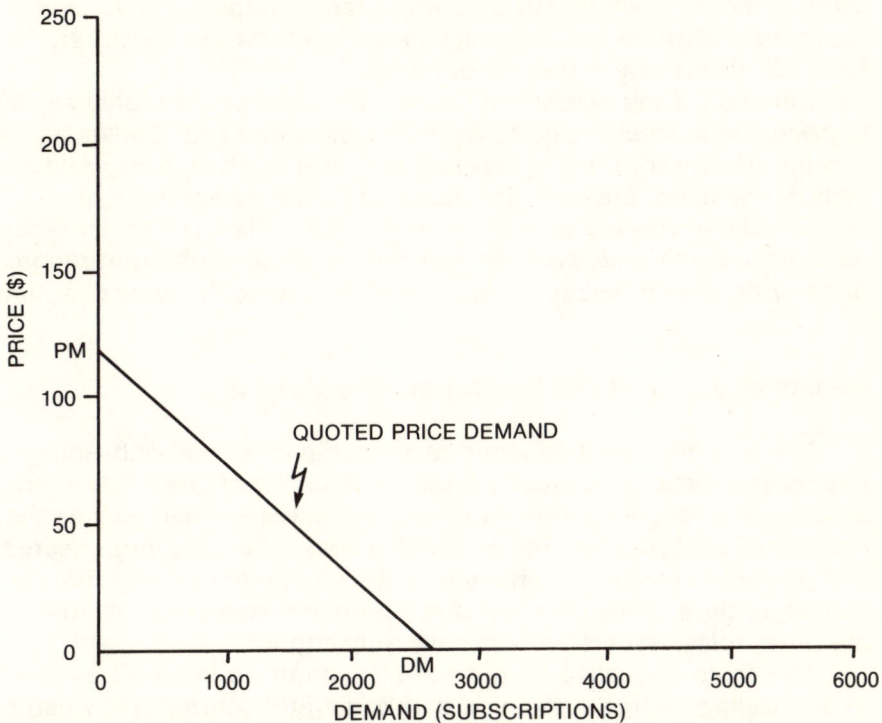

Figure 8.6 Price and demand (subscriptions) relationship for
library subscriptions.

Fourth, greater awareness of journals generally will shift the
demand curve to the right. With more people aware of a journal, there
are likely to be more subscriptions to that journal at a given price.

Finally, the shape of the curve is certainly not linear. However, to
our knowledge no one has been able to estimate the actual price
demand curve, if indeed such a curve actually exists over all ranges of
demand. The problem with existence is delicate, since all but a single
point on a curve are hypothetical. Even with these problems, however, it
is useful to understand the economic principles associated with
publishing scholarly journals.

Two recently performed studies deal with the relationship between
price and the number of subscriptions. Both apply a multiattribute
analysis approach to determine the relative contribution made by price
and other attributes such as editorial quality, graphic quality, and so on
to number of subscriptions and use by individuals. The first of these
studies, performed by David Evans of Charles River Associates,[9] applies
a variation of a multiple-regression modeling approach. Dennis
McDonald of King Research,[10] on the other hand, used a simple linear

additive model in which attribute importance weights (for example, importance of price, journal prestige, and publication speed) are assessed directly by survey respondents.

Generally if the number of journal subscriptions is highly sensitive to price, the journal is said to have an elastic demand. Conversely if the number of subscriptions is scarcely sensitive to price, a journal has an inelastic demand. Elasticity is measured by the comparison of changes in quantitiy associated with changes in price.* Elasticity varies from journal to journal and even for a given journal over different ranges of price or number of subscriptions.† And, of course, it varies over time.

Relationship of Cost and Number of Subscriptions

The second basic economic relationship in journal publishing is the relationship between cost and number of subscriptions.‡ The average production cost per journal subscription decreases markedly as the number of subscriptions increases. The reason for this large decrease is that journal publishing is characterized by large fixed costs that do not vary regardless of the number of subscriptions and relatively low marginal costs associated with each subscription.

The costs discussed here deal only with those incurred by journals in processing, reproducing, and distributing the journal. They can be aggregated into three groups:

1. Costs of processing information (prerun): Examples include editing, redacting, and preparing master images.
2. Costs of reproducing and distributing journals (runoff): Examples include processing an order, reproducing a single copy, mailing individual copies, and maintaining mailing lists.
3. Indirect costs: There are other cost elements such as rent, overhead, and administration that are not sensitive to the number of copies produced.§

*Price elasticity, $EP = -(\Delta q/q)/(\Delta p/p)$. If the absolute value of Ep is relatively less than 1, demand is relatively inelastic and total revenue would increase with an increase in price.
†It is reasonable to postulate that there is a very low elasticity in the vicinity of a base demand, D_M, which represents the subscriptions by libraries attempting complete coverage in their specialties, and a very high elasticity near P_M.
‡Cost curves rather than the traditional supply curves of producers in a competitive economy are fundamental to our discussion for two reasons: (1) scientific and technical journals are highly differentiated goods (one journal does not readily substitute for another) and (2) publishers generally are operating in a region where there are not decreasing returns to scale.
§Capital investment is a relatively minor factor of production because substantial working capital derives from advance subscription payments.

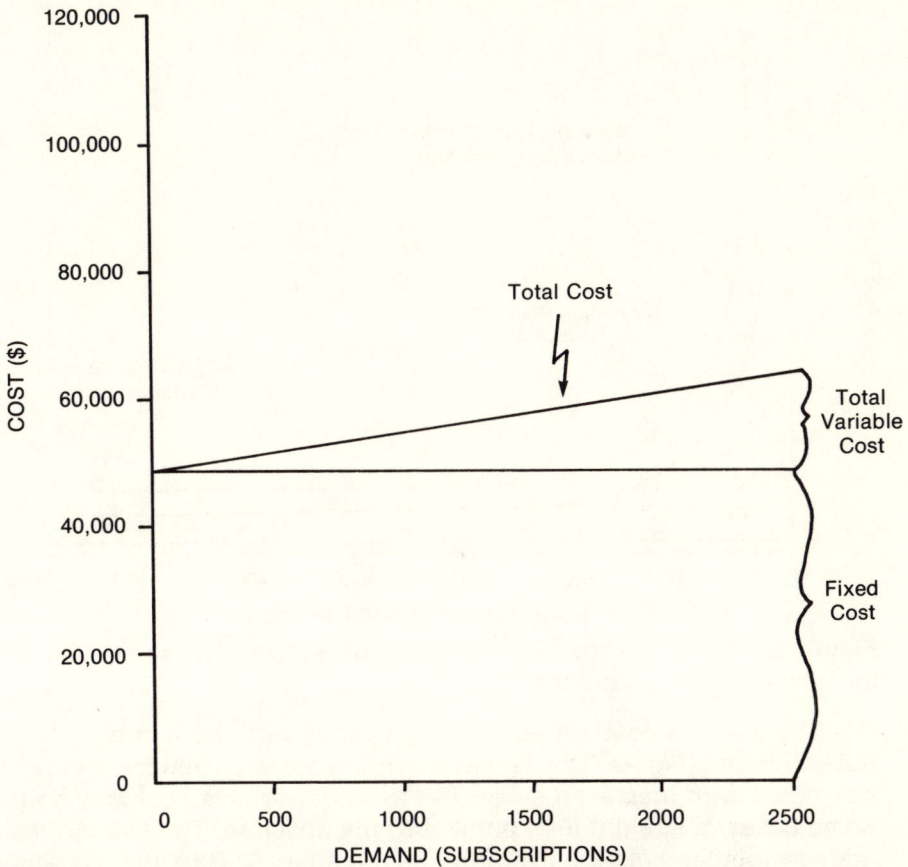

Figure 8.7   Cost and demand (subscriptions) relationship.

The costs of publishing also vary by such journal characteristics as the number of issues published per year, articles, pages, characters per page, and graphics and by the cost rates of labor, equipment, supplies, postage, and so on (all of these are included in the publishing cost model described in Chapter 4). Applying this model to the average journal characteristics and cost rates observed in 1977 yields the relationship between cost and the number of subscriptions shown in Figure 8.7.

The total variable costs, fixed costs, and total costs are plotted in Figure 8.7. The fixed costs include the direct prerun costs and fixed costs such as overhead. Variable costs include the direct costs attributable to reproduction and distribution of journals. The integral of marginal costs, for each successive unit (copy) from the first to the nth, is the total variable cost.

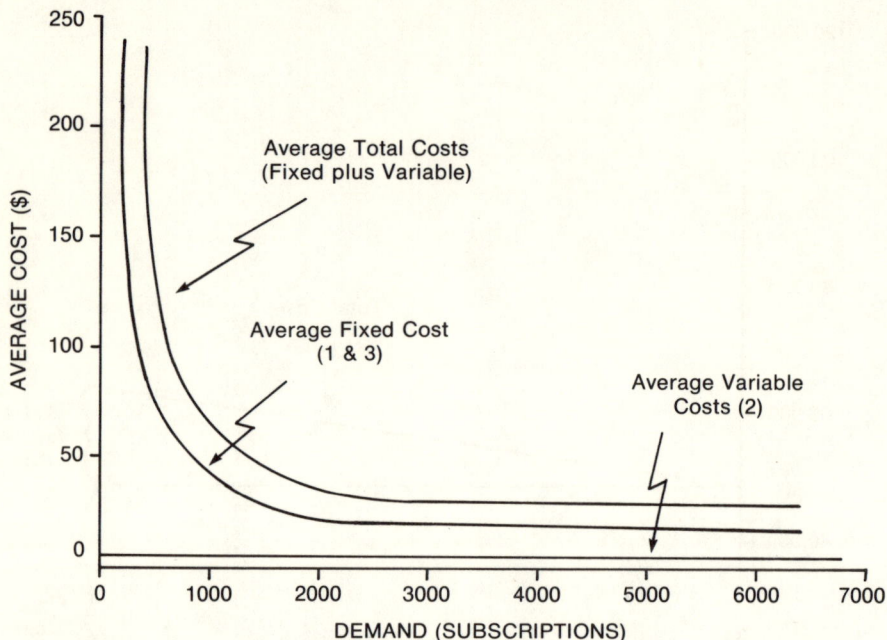

Figure 8.8 Average cost and quantity (subscriptions) relationship for individual subscriptions.

The average cost per subscription varies with the number of subscriptions (Figure 8.8). The average fixed cost produces a curve that decreases with increased sales. The average variable cost may be (in some cases) a straight line, parallel to the abscissa. The two curves added together yield the average total cost curve. Thus the average total cost decreases as sales increase and approaches the average variable cost at large numbers of subscriptions. Various alternative pricing policies take into account the relationships of average cost, average variable cost, unit prices, and social value.

Alternative Journal Pricing Policies

Journal pricing policies are usually determined by the mission, goals, and objectives of the journal publishers. Commercial publishers must make a profit on their journals to achieve a return on their investment and to obtain a return for the risk associated with publishing the journal. Society publishers, on the other hand, may attempt merely to recover costs so that they do not lose money on the journal. Government publishers and some society publishers may wish to

Table 8.11   Estimated cost-quantity and price-quantity demanded relationships for individual and library subscriptions.

Quantity Demanded	Individual Subscriptions			Library Subscriptions		
	Cost	Price	Profit	Cost	Price	Profit
250	$198	$85	($28,250)	$122	$108	($3,500)
500	102	80	(11,000)	64	95	15,000
1,000	54	71	17,000	35	70	35,000
1,500	38	62	36,000	26	45	28,000
2,000	30	52	44,000	21	20	(2,000)
2,500	25	43	45,000	18	5	(32,500)
3,000	22	33	33,000	16	0	(48,000)
3,500	20	23	10,500	15	0	(52,500)
4,000	18	14	(16,000)	14	0	(56,000)
4,500	17	5	(54,000)	13	0	(58,500)
5,000	16	5	(55,000)	12	0	(60,000)
5,500	15	0	(82,500)	12	0	(66,000)
6,000	14	0	(84,000)	11	0	(66,000)
6,500	14	0	(91,000)	11	0	(71,500)
7,000	13	0	(91,000)	10	0	(70,000)
7,500	13	0	(97,500)	10	0	(75,000)
8,000	12	0	(96,000)	10	0	(80,000)

Note: Figures in parentheses represent a loss.

recover only a portion of their costs and to subsidize the remaining portion in order to achieve greater social benefit from their journals.

The differences among the three pricing policies can best be shown by combining the demand and cost curves. These two curves are superimposed in Figure 8.9. There are two break-even points where the two curves cross. The prices at these two points are designated by P_{BE} and P'_{BE}. At prices above P'_{BE} the cost curve is above the price curve so that a loss would be incurred. At all prices between the two break-even points (P_{BE} and P'_{BE}) the demand curve exceeds the cost curve so that a profit* is made at any price in that range. At prices below the break-even point P_{BE} a loss is incurred because the cost curve again exceeds the demand curve. The amount of the profit or loss can be calculated by multiplying the number of subscriptions at any given price by the difference between the price and the cost for that number of subscriptions. The price, cost, and net revenue (profit or loss) are displayed in Table 8.11 by a range of number of individual and library subscriptions. The price-demand relationship is hypothetical, but the

*In some instances, the excess net revenue is applied by a society publisher for research, equipment investment, or other society services so that a profit in a tax sense is not made by the publisher.

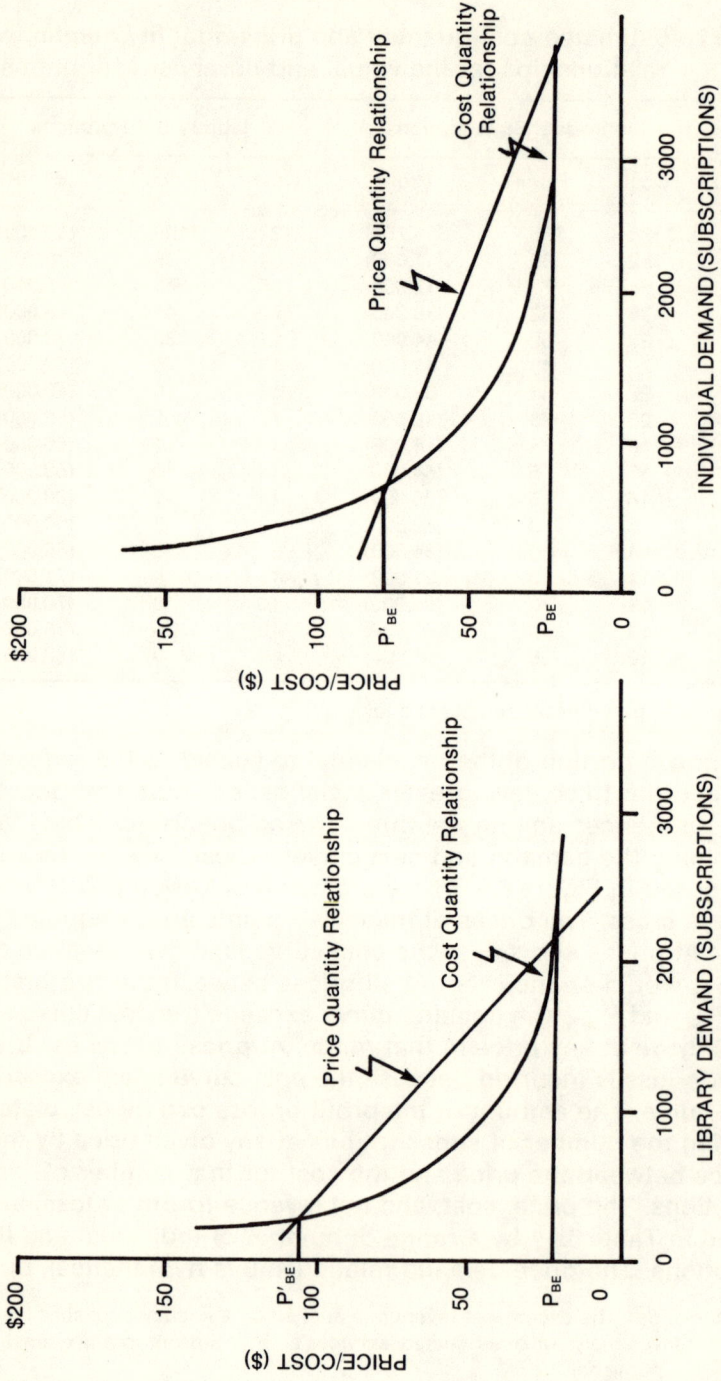

Figure 8.9 Average cost, price, and quantity (subscription) relationships.

236

costs are computed from the cost model described in Chapter 4 using average journal characteristics.

The average journal publisher would lose money on individual subscriptions at all prices above the upper break-even point of $78. There would be a profit at all prices between $78 and $19, and a loss would be incurred at all prices below the lower break-even point of $19. For library subscriptions, a profit could be incurred between the prices of $105 and $21.

Commercial publishers normally like to maximize profit. The maximum profit for the average journal occurs at prices of about $43 and $70 for individual and library subscriptions, respectively. Ordover and Baumol demonstrated that publishers should substantially increase library subscription prices if they wish to maximize profits.[11] These publishers in particular have sources of income beyond subscription fees from their publications, such as advertising, sale of reprints, and sale of microform.* In 1977, we estimated an average of $11,700 for advertising. Potential advertising income is, however, highly dependent on the number of subscriptions. A rule of thumb is that advertisers will pay about $40 per page of advertising for each 1,000 subscribers.[12] With very large subscription journals, say 10,000 subscribers, the amount comes down to about $10 per 1,000 subscribers. Journals with a small number of subscriptions often have no advertising because of the small number of readers reached. The average revenue from advertising in 1977 is estimated to be $2,350 for journals with fewer than 3,000 subscribers, $11,750 for those with 3,000 to 10,000 subscribers, and $47,000 for journals with over 10,000 subscribers. Another factor is that marketing for advertising is quite expensive—about $0.50 to $0.60 for every dollar of revenue, much of it due to fixed costs. When runoff costs of the advertising page are added, it is necessary to have nearly 50 percent of the space dedicated to advertising to make it economically worthwhile to the publisher.

Most society and nonprofit publishers attempt to achieve a break-even price. This can be done when all costs (fixed and variable) are recovered through subscription revenue and other sources of income. Economists refer to full cost recovery as average cost pricing. It is difficult to achieve an exact break-even even though publishers know the approximate number of subscriptions ahead of time.

In some instances a publisher will attempt only to recover the cost of reproducing and distributing additional subscriptions, a policy referred to by economists as marginal cost pricing. This pricing policy is employed by some government journal publishers where the prerun

*Some publishers receive royalties for authorized photocopying, but this is at best a minor source of income.

costs are absorbed by the government and by some publishers that receive income from authors by page charges, which cover a portion or all of the prerun costs. The practice of requiring author page charges began in the 1930s in the Physical Sciences and is now prevalent in that field as well as in Engineering. Page charges are almost nonexistent in the Social Sciences. Commercial publishers rarely require page charges, partly because the federal government does not allow them as a direct cost on a research grant or contract if the results are reported in a profit-making publication. It is clear that the prices for journals will drop substantially if the prices recover only the runoff costs.

THE SOCIAL BENEFIT OF SCIENTIFIC AND TECHNICAL JOURNALS

Some understanding of the social benefit of scientific and technical journals is important so that long-term planning by system participants can be made in light of the social consequences of their decisions. At one extreme it might be argued that scientific and technical journals are a pure public good, in which case government funding might be considered appropriate. At the other extreme an argument is made that these journals are like all other goods, and economics should be entirely determined by the free market.

An excellent discussion of the public good properties of scientific and technical journals is given by Ordover and Baumol.[13] They define a pure public good as a good or service in which total cost is completely unaffected by the number of persons served. Standard examples of pure public goods are national defense and scientific discovery. Two important attributes of pure public goods make government financing appropriate. The first is that there is zero marginal cost for additional users. This implies that a charge for the consumption of such a good or service is undesirable. The second attribute, referred to as nonexcludability, is that it is impossible to deny any person the use of a pure public good, which implies that a charge is impractical. Thus government (or charitable) financing would be the only appropriate source for a pure public good.

Even if scientific discovery* is a pure public good, the question remains whether its medium of communication should also be considered a pure public good to be financed by government. The marginal costs associated with printing, distribution, acquisition,

*For the sake of simplicity, we will consider here only scientific discovery resulting from research financed by the government. In that way we largely avoid the issue of proprietary rights.

storage, and use of journal articles are far from zero. Thus even though marginal costs per use are very low compared to the cost of the research that generates information, journals cannot be considered a pure public good. Ordover and Baumol reach that conclusion and then concentrate their discussion on establishing optimal financial arrangements. They conclude, "The standard argument advocating free provision of public goods does not hold in the journal case since marginal cost is in fact positive. Supply of any good at zero price always leads to wasteful use of resources if the resources cost of its increased use is nonzero."

At the other end of the spectrum one could argue that scientific and technical journals are like any other private good and should be able to withstand the test of the marketplace.[14] The principal weakness of this argument is that the cost structure of publishing is such that a substantial number of subscriptions are necessary to sustain a journal. However, there are many journal articles of high scientific quality that are of direct use only to a small number of persons. Furthermore these articles may be used over a long period of time and must be stored for future use in some form or another. If journal publishers relied exclusively on revenue from distributing these articles, many such journals would not survive. Instead society publishers, in particular, rely on subsidies by means of bundling the journals with other benefits or services such as membership in the society. Scientific and technical journals are not a pure public good, but they also should not reside in a pure market environment because some social benefit would be lost.

Ultimate social benefit from information found in journals can be derived from such effects as possible improvement in the nation's health, or energy resources, or economy. Unfortunately it is virtually impossible to measure the contribution journals make to social benefit. Nevertheless persons have attempted to apply economic principles and to make some gross estimates of the social value of scientific and technical journals as perceived by journal users. The approach assumes that a scientist's perception of value is at least correlated with social benefit; then the social value can be used as surrogate measure of social benefit.

The Value of Journals to Society

Journals published in the United States contain substantial information that must be recognized as an important national resource. One popular economic theory concerning a partial estimate of the value of scientific and technical journals is based on the effective price that

240

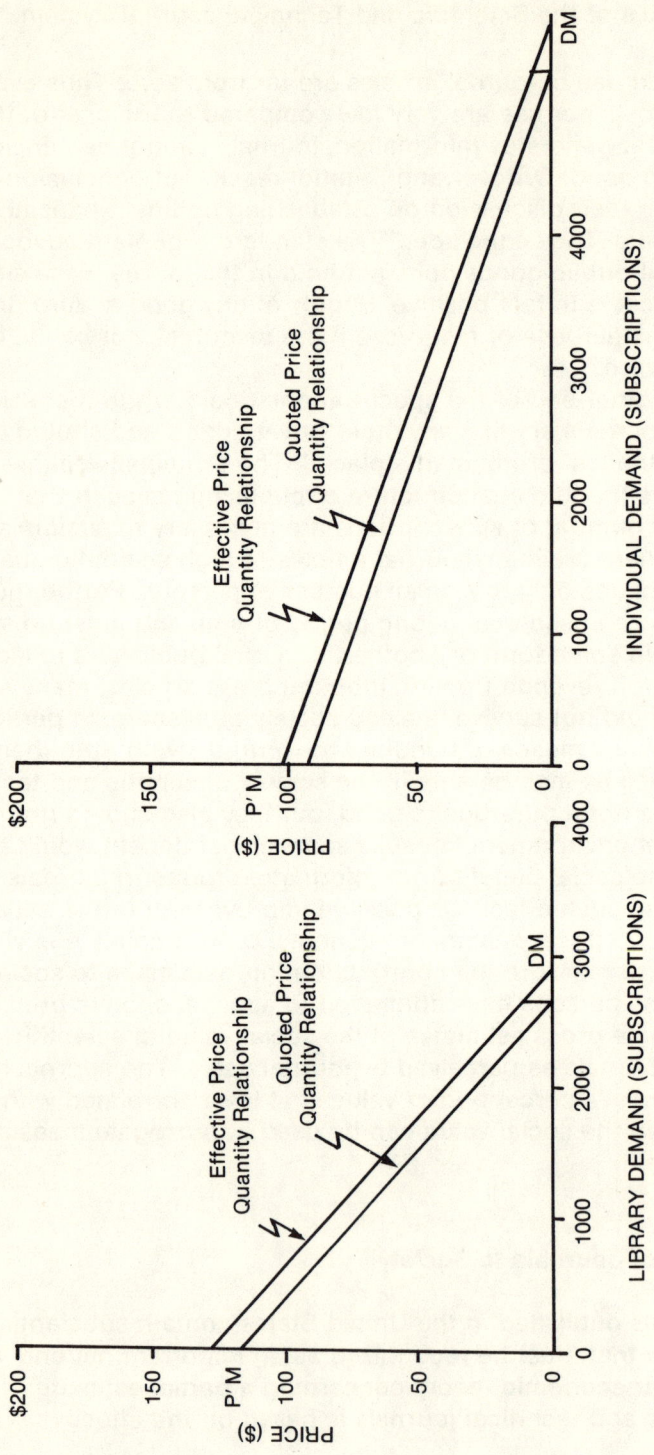

Figure 8.10 Price and quantity (subscriptions) relationships with user cost added.

users are willing to pay for information.[15,16,17] The effective price under this approach is the subscription price plus the price paid by users in their costs associated with identifying, locating, ordering, receiving, using, and storing scientific and technical journals. The approach is that the partial contribution to social benefit made by each scientist or engineer due to the information is the same as the self-interest value perceived by that user. The value realized from providing a user with information is then measured by the maximum effective price the scientist or engineer is willing to pay for it. This price is partially measured by the subscription price demand curve shown in Figure 8.5. If user costs are added to subscription price, one arrives at a curve such as that given in Figure 8.10* It is significant that the total effective price is higher than the quoted subscription price at all levels of demand. Thus even at the maximum demand (D_M), where the subscription is free, there is still a cost paid by the user; the effective price is always greater than zero.

Under this approach, the total value of a journal, if it reaches all potential subscribers, is provided by the entire area under the demand curve. Thus it is thought that one can establish, in principle, the effect that different price levels have on the value derived by society through analysis of the area under the curve. For example, assume that a subscription price is P_1 and that the total effective price (including cost to user) is P'_1. Then the number of subscriptions sold (D_1) is given as shown in Figure 8.11, as the quantity at which the effective price demand curve intersects the price P'_1. The total value received by all users who purchase the journal is thought to be the area under the effective demand curve to the left of the vertical line at D_1. This area is shaded in Figure 8.12.

This assessment can be better understood if the area under the curve is subdivided into two parts: the rectangle formed by the effective price (P'_1) times the number of sales (D_1), and the area above the rectangle. The rectangle represents the effective price times the sales at that price, one measure of the value of the journal at that particular price. However, some users are willing to pay more than that price, which means that they place a value on the journal higher than what they had to pay for it. This excess value is represented by the area above the rectangle. (However, it is disputed whether this area can realistically be included in an assessment of value.)

The net social value of the journal is the total value derived minus the total cost. The total social value is that which is actually received by users. Thus the remaining area under the curve (to the right of the

*For simplicity, a constant added cost per subscriber is assumed, independent of subscription price.

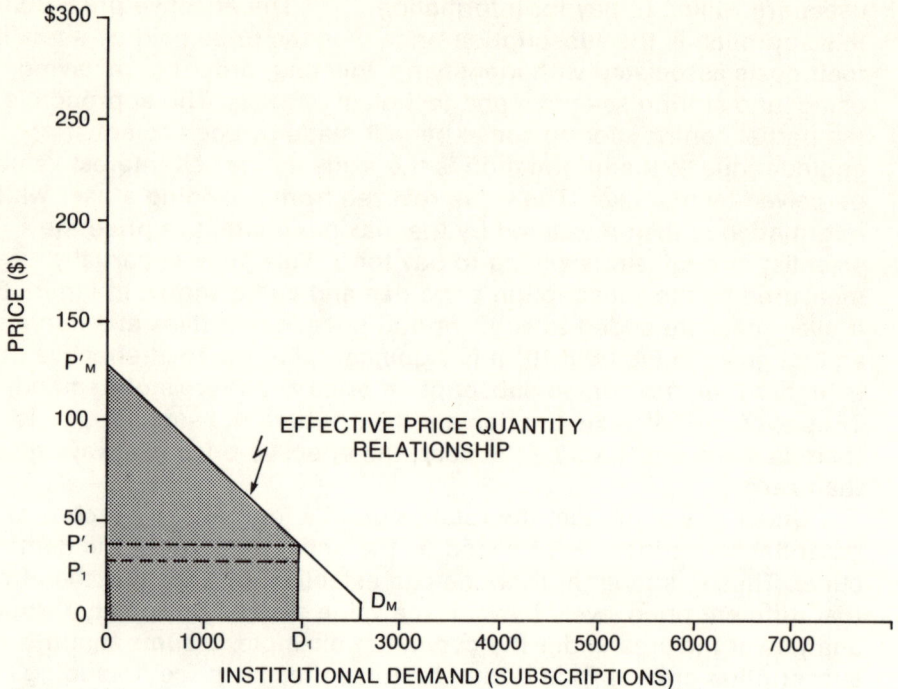

Figure 8.11 Area under the curve of the price and quantity (institutional subscriptions) relationship with user cost added.

vertical line intersecting the effective demand curve at P'_1) represents the value to society that is lost because users are required to pay P_1 for the journal. If the subscription price were zero, presumably all needed subscriptions would be ordered and the total social value would be increased. Conversely if the price of the journal is increased, the value to society is decreased.

This discussion is purposely oversimplified. First, it assumes that there is no substitute for the journal subscription. In fact, a scientist who is unwilling to pay for the journal can go to a library or request a copy of a needed article from the authors. Also one can purchase some other information product or service that is considered to be of equivalent value. Second, if we considered this approach with all products and services, the total value of them would far exceed a reasonable amount. Therefore it is not appropriate to employ this approach in estimating the value of information products and services. It is relatively safe to assume, however, that small changes in price would alter the demand and that this incremental change would affect the social value in the same direction. What cannot be measured is the amount of that change in social value.[18]

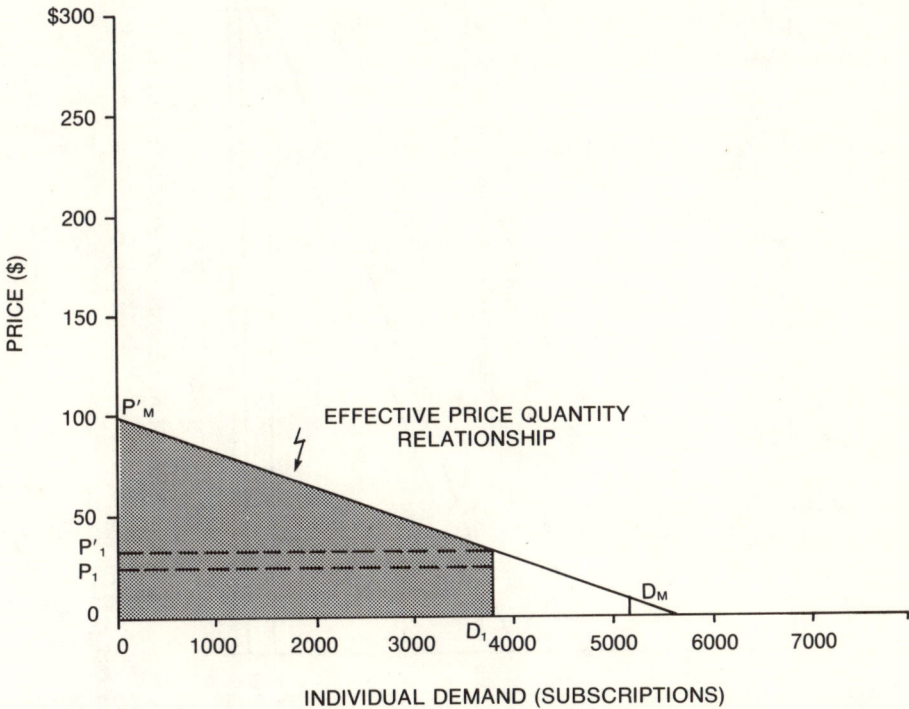

Figure 8.12   Area under the curve of the price and quantity (individual subscriptions) relationship with user cost added.

AN ASSESSMENT OF ALTERNATIVE SUBSCRIPTION PRICING POLICIES

For the most part, pricing policies depend on the return on investment and the amount of cost that publishers would like to recover for processing and distributing their journals, as well as the net social value achieved in some cases. The latter objective depends partially on the mission of the publishers and the social benefit imputed to their journals. Conceivably some journals produced, say by the government, may be of such high social benefit that they should be given away free in order to encourage their use. Charging for such a journal would only create additional, unwarranted costs for accounting, billing, and the like. These circumstances rarely occur, so publishers usually decide to charge some amount to recover costs. The principal ways in which this is done is to charge only for the variable costs, for full costs,* for a point

*Publication costs only, exclusive of the costs of generating the information published.

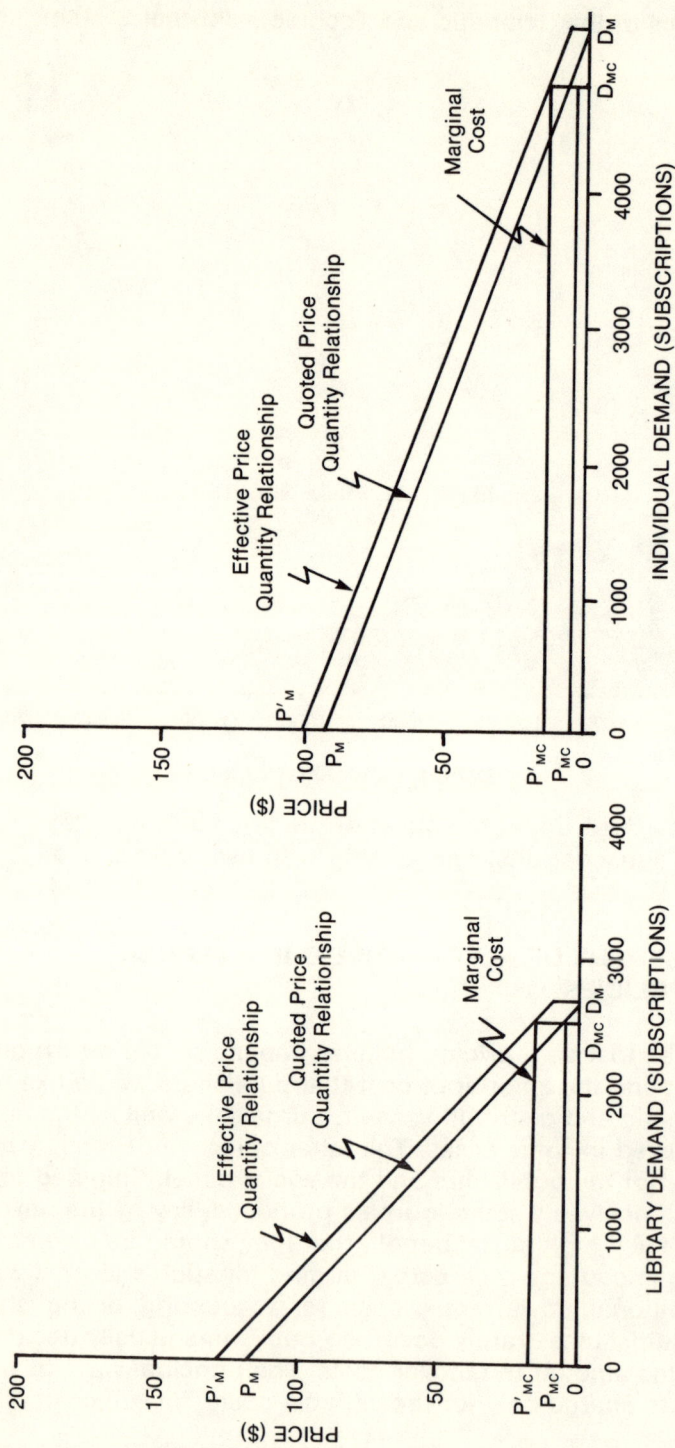

Figure 8.13 Price quantity cost curves for marginal cost pricing.

244

between these, or to optimize the revenue. Variations of these policies and other pricing policies include price discrimination to certain user communities, charging what is considered a fair market value, pricing to accomplish a desired objective, and pricing when more than a single product is produced from a common source.

No Charge

The question arises why not to give journals away free and gain the entire social value represented by the demand curve, with the journal financed by the government. The principal argument against this is that there may be frivolous requests for the journal. If journals are given free, some scientists (or libraries) may acquire unneeded subscriptions. However, even free journals cost users something in terms of their time to identify, locate, order, receive, use, and store copies. Also the maximum net social value may not be achieved at zero price. The net social value in this case is the total social value less costs to the publisher and to the users. In fact the optimum net social value will probably be achieved when the price is not free but is closer to the marginal cost.

Each additional subscription has a value equal to the price one additional subscriber is willing to pay. For quantities up to the point D_{MC} at which the value placed on the last subscription equals the marginal cost (C_M) of supplying that subscription, the social value added by each subscription exceeds the added cost. At higher quantities, the value ascribed to an additional subscription is less than the added cost of providing it; hence additional subscriptions would decrease the net social value.

Providing subscriptions free therefore makes sense only in two situations: if social benefit is deemed well served regardless of social value or if the additional cost of administering subscription fees would exceed the net social value lost by providing low-value subscriptions.

Marginal Cost Pricing

Marginal cost pricing occurs when a publisher recovers only the cost of reproducing copies, handling requests, and distributing each additional subscription. Thus there must be some subsidy from a professional society, from authors (perhaps through page charges), or from the government to recover fixed or prerun costs. The demand and cost curves for marginal cost pricing are given in Figure 8.13. The

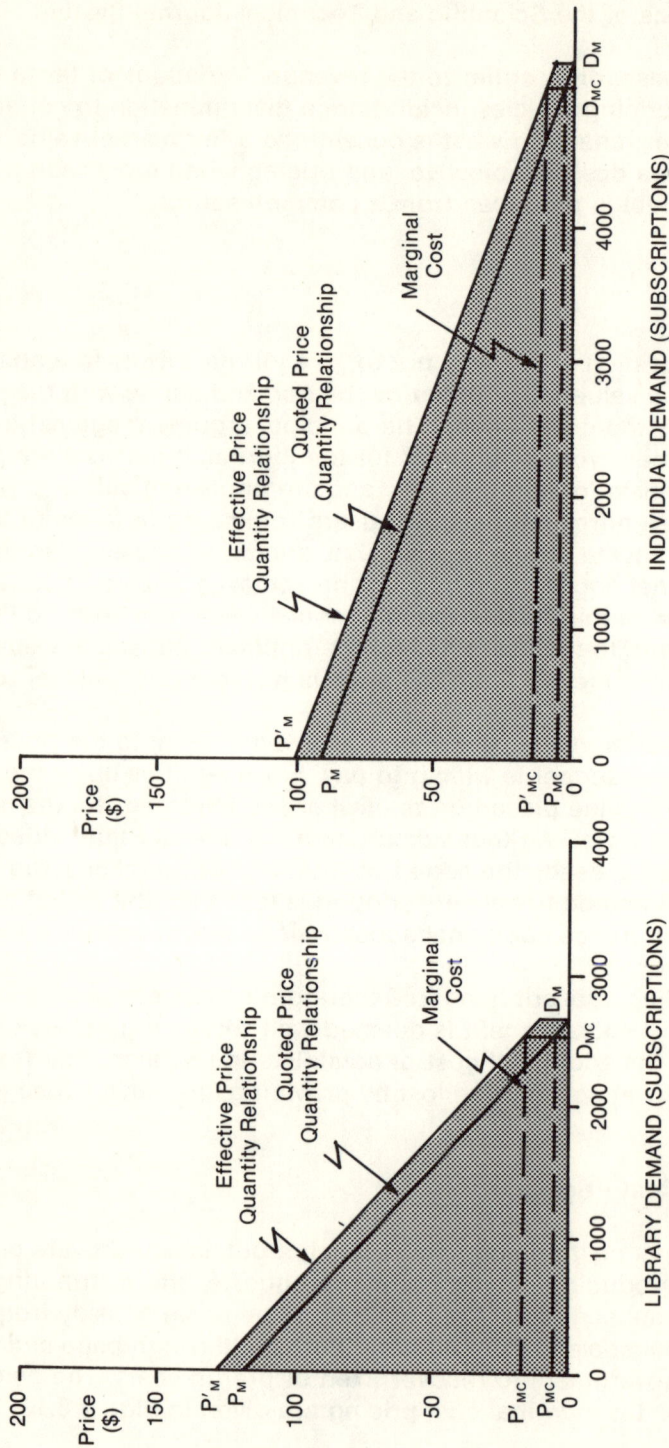

Figure 8.14 Area under the curve of the price quality cost curves for marginal cost pricing.

cost curve has been drawn to show constant marginal cost. This is a reasonable approximation. Typically the marginal costs of journals are very low, so that very little value would probably be lost to society by charging this amount. Since the marginal cost reflects the actual resources used to produce one additional unit, a marginal cost pricing policy is usually considered by economists to be the policy that maximizes net social value.[19] There is an exception to this general rule when there are economies of scale and high fixed costs. In this situation, it may be preferable (in terms of net social value) to set prices slightly above marginal costs so that all costs are recovered.[20]

The price that would be chosen under a marginal cost pricing strategy is the one for which the marginal cost equals the unit revenue. At this price, the value of the journal is shown by the shaded area in Figure 8.14. The total marginal cost and the total revenue to the publisher is given by the cross-hatched rectangle.

Most marginal costs are attributable to reproduction, distribution, and subscription handling, but even here some fixed costs may be hard to distinguish. For example, on a printing press run, there are fixed costs associated with setting up and shutting down the presses. Also a publisher tries to anticipate the total demand and prints a number of copies beyond subscriptions so that a fixed storage cost is necessary and some copies may remain unsold. In some instances, indirect costs such as rent and utilities are allocated on a unit cost basis. Thus the cost curve usually would not be a straight line, parallel to the abscissa, but rather a curve slightly concave upward.

Average Cost Pricing (Full Cost Recovery)

Average cost pricing occurs when a publisher decides to recover all of the costs associated with processing and distributing its journals. The average cost pricing policy may yield some less value to society than the marginal cost pricing policy since the price will always be greater. It is possible that cost is greater than the subscription price over all levels of demand. This does not necessarily mean that a positive net benefit to society would not be achieved. However, publishers would never be able to recover full costs at any level of subscription under this circumstance. The social value lost by using average cost instead of marginal cost pricing is represented by some amount close to the area under the effective price demand curve between D_{AC} and D_{MC} in Figure 8.15; the net social value lost is represented by that portion of this area above P'_{MC}. The revenue (and the total average cost) is shown by the cross-hatched area. Note that P_{AC} is the same as P_{BE}, the lower break-even price, in Figure 8.9.

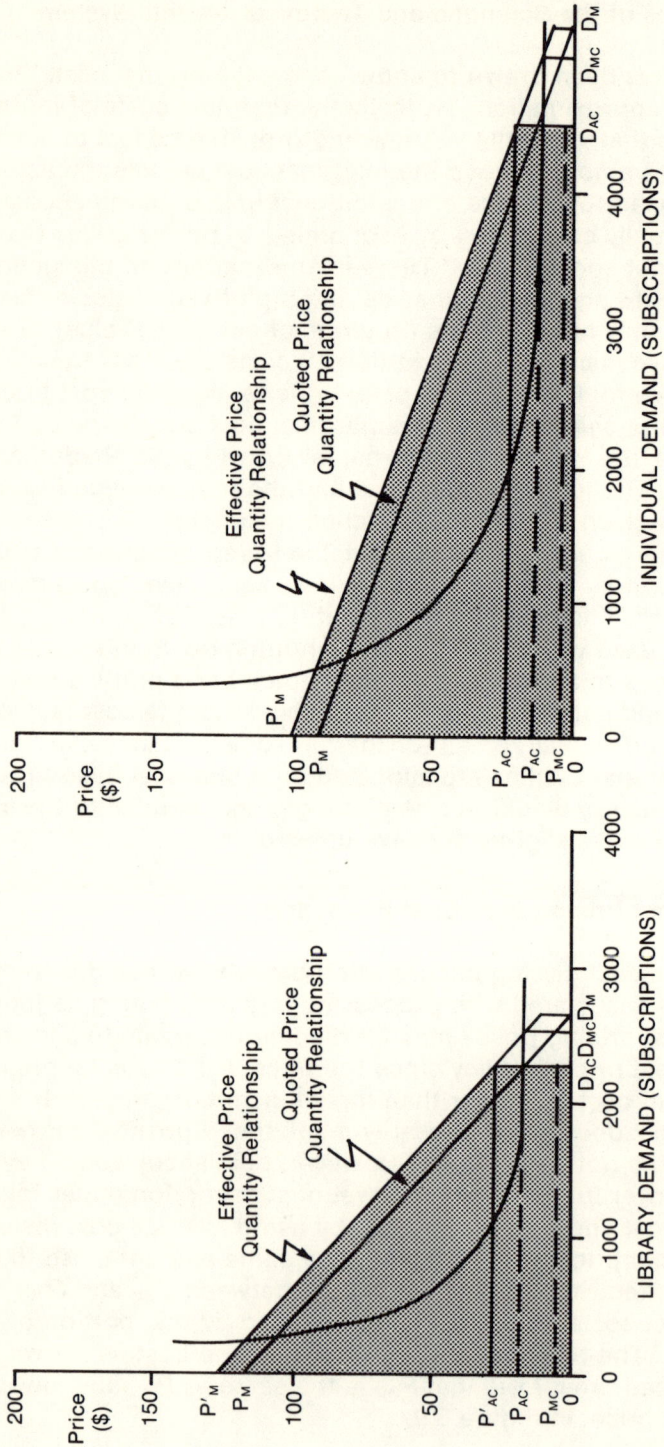

Figure 8.15 Price quantity cost curves for average cost pricing.

248

One problem with average cost pricing is that it is very difficult to establish the break-even point in the absence of any knowledge of the price demand curve. This is particularly true with journals that have a high fixed cost and thus a widely varying average cost. Here the risk may be very great, and a large loss may be incurred or unwanted profit made. With low fixed costs or with marginal cost pricing, the risk is not as great because the choice of prices can be made from a relatively small range.

Optimum Profit Pricing Policy

Most commercial publishers attempt to establish prices that will optimize their profits. A profit is made at any time the revenue exceeds cost, which occurs over a range of demand shown in Figure 8.9 above P_{BE} but below P'_{BE}. Economic theory indicates that the optimum profit can be achieved at the price at which marginal cost equals marginal revenue. The price at which optimum profit is achieved will probably yield less social value than average cost pricing.

Other Pricing Policies and Considerations

One potential pricing policy is to charge what is considered to be a fair market value (a price that considers the worth of the journal and what others charge for similar publications). The problem with this policy is that the unknowns and risks are very great unless substantial experience has been gained in the marketplace. Thus it becomes difficult for most publishers to budget for either excessive or inadequate demand that may occur at fair market value price. Since it takes no explicit cognizance of cost or demand, it may be equivalent to any or none of the last three pricing policies.

Price discrimination is used by some publishers where some user groups are charged differently from others.* The most obvious way to accomplish this is to charge groups of users based on their price elasticity. For example, library subscribers may have a less elastic demand than individual subscribers; thus they are sometimes charged more. Also U.S. subscribers may be charged less than non-U.S. subscribers. There are other purposes for price discrimination as well,

---

*On July 31, 1978, the Internal Revenue Service held that professional societies are not allowed to give a preferential subscription rate to members or collect page charges and that advertising revenue to professional journals is taxable income. If this ruling stands, it drastically alters the entire financial structure of journal publishing.

such as to develop loyalty, to take advantage of favorable price elasticity for some groups, and so on.

Allocation of costs between multiple products from the same editorial inputs may provide a lower cost basis for a journal. Pricing by marginal cost, average cost, or maximum profit would then result in lower subscription prices (and higher social value) than the corresponding policy applied to the total cost.

Sometimes prices are established to accomplish an objective. For example, a price may be purposely set low to encourage use of an information product or service that might not otherwise be used. Another example is when more than one product is produced from a common source, such as a data base. For example, the National Technical Information Service, which sells technical reports in paper and microform, has substantially increased the price of paper reports and maintained a low price for microform. As a result, the proportion of on-demand requests for reports shifted dramatically from paper to microform. Finally a publisher may decide that the price for announcement products, search services, and the like should be kept below cost if these services create an increased demand for primary information. This is particularly important if demand for the primary product is elastic. Increased awareness of the information may thus provide higher net social value than would a reduction in price. This is even more true if a secondary product or service produces a reduction in the effective cost to a user by lowering additional costs.

ECONOMIC INTERFACE BETWEEN AUTHORS AND PUBLISHERS

Very few publishers of scientific and technical journals compensate authors for their manuscripts or pay them royalties. Maybe in the future publishers will compensate authors for providing machine-readable manuscripts such as those produced on word-processing equipment since this could drastically reduce prerun costs. To our knowledge this is not done with journals, although it is with book manuscripts. On the other hand, some journals charge authors in the form of page charges or by some other means.

There is a demand curve for such payments just like that observed for other information products or services. As the charge (price) increases, fewer manuscripts will be submitted to a journal. Presumably there is some page charge at which no manuscripts would be submitted and some maximum number of manuscripts that will not be exceeded even with no charges. Between these two points is a demand curve of some unknown shape. Two studies have attempted to shed some light on this relationship but without strong evidence.[21] If page charges are paid, it is assumed that the price of a journal will be reduced, thereby yielding a greater quantity demanded (and hence some increase in social value). It can be argued, however, that price is not lowered or at least not lowered equivalently by publishers. A good analysis of this page charge, cost, and subscription price relationship is given by Braunstein.[22]

A policy question currently in dispute is whether the page charges for articles reporting the results of federally sponsored research should be paid out of the grants or contract funds.[23] The present Federal Council for Science and Technology (FCST) policy states that payment of page charges is recommended under the following conditions: research papers report work supported by the government; charges are levied impartially on all research papers published by the journal; payment of such charges is in no sense a condition for acceptance of manuscripts by the journal; and the journals involved are not operated for profit. One extreme possibility is that page charges be forbidden as a direct cost to a grant or contract, and the other extreme is that every grantee or contractor be required to include page charges as a direct cost. Unfortunately there currently is no economic basis upon which to make such a policy decision.

The basic trade-off for federal government payment of page charges is that net publishing costs would drop, resulting in reduced prices to subscribers. If this happens, there should be more subscribers to journals, and more social value presumably would be derived from the journals. A hypothetical example illustrates generally what one could expect in terms of social value by applying the two extremes. The example is given for the Physical Sciences, using an economic demand curve derived by Berg[24] for individual and institutional subscribers to some chemical journals. For the sake of simplicity, we have used the more conservative of the two models (individual). The model, in effect, says that the number of subscribers is a function of price, number of pages in the journal, and number of scientists in the field:

$$\text{Subscriptions} = 6{,}370 + (.061 \times \text{pages}) - (39.6 \times \text{price}).$$

In the Physical Sciences, the average number of pages is 1,240 (207 articles), and the average price weighted by circulation is $38.89. Applying Berg's equation for chemical journals to all Physical Sciences thus yields an average number of subscriptions of 4,906 (included are domestic and foreign subscribers).

There is an economic demand curve for authors and their articles as well.[25] Here we assume that eliminating page charges will increase the number of pages submitted to journals from 1,240 to about 2,000 pages (about 333 articles). Similarly requiring page charges on all articles should decrease the number of articles submitted to about 800 pages (about 133 articles). Nothing is assumed about the quality of articles not submitted. We do assume that the length of articles submitted remains constant at 6 pages per article.

A simplified form of a cost model for Physical Sciences,* for numbers of pages and of subscriptions near their average values, is:

$$\text{Cost} = \$18,167 + (\$94.12 \times N_{pages}) + (\$14.19 \times N_{subscriptions}).$$

Since the average number of pages is 1,240 and the average number of subscriptions is 4,906, the average cost comes to $204,492. Average revenue comes to $195,700 ($38.89 × 4,906), so there would appear to be a loss on current journals. However, an estimated average of about $36,000 is now derived from page charges. Subtracting this number from costs yields an average excess (profit) of about $4.50 per subscription. If the number of pages increases to 2,000 or decreases to 800, there will be a substantial effect on cost. Also, covering the $94.12 prerun cost by page charges will have a significant effect on cost, price, and number of subscribers. Cost curves are estimated for the following three conditions: (1) no page charges will be paid, (2) page charges will be paid for all articles, and (3) page charges will be paid only for articles reporting federally sponsored research. The demand curve is computed only at the current average number of pages since a difference between 800 and 2,000 pages scarcely affects the demand and has an inconsequential effect on social value.

The cost curve when no page charges are paid, is just below the demand curve. Thus at no point will a profit be over $4.50 per subscription. The point of maximum profit appears to be a price of about $89 per subscription and a circulation of 3,000 subscriptions. A comparable point is established for the cost curve when all articles require page charges and when an average profit of $4.50 per subscription (the average now achieved) is made. This point is at a price of about $21 and a circulation of about 5,800. Thus some added but unknown amount of social value should be achieved by having all prerun costs covered by page charges.

Finally, we assume that all articles reporting federally sponsored research will be covered by page charges. Again a $4.50 average profit per subscription is assumed. The appropriate price is then about $35, and the number of subscribers is about 5,250. Thus the social value with this condition should also be greater. The cost to the federal

*The constant factor and effect of this field of science is combined into one term.

government for purchasing the prerun services is $138,000, so that the social value achieved from such payment of page charges would have to be substantial.

No attempt is made in any of these analyses to adjust for income from foreign subscriptions or to reduce the national social value by the added value accrued to these subscriptions. This analysis is gross, but it does serve as a framework for studying the effect of page charges and what benefit might be gained if the federal government paid them.

FEDERAL CONTRIBUTION TO SCIENTIFIC AND TECHNICAL COMMUNICATION

The federal government is engaged at various levels in each of the eleven principal scientific and technical communication functions. It makes substantial contributions to research and development funding in the generation function; its employees write articles and many other authors are funded under grant or contract; it publishes hundreds of periodicals and thousands of scientific and technical reports and patent documents each year; it has over 3,000 libraries and clearinghouses and funds other libraries through grants; it operates or contracts out for numerous abstracting and indexing services; and a large portion of journal readers are federal government employees or are funded under grant or contract.

In order to determine the magnitude of these federal government expenditures, we conducted four surveys. The first two involved authors of scientific and technical articles and government-sponsored technical reports. From these we determined the proportion of articles that reported federally funded research and the amount of expenditures involved in preparing journal articles and technical reports. Another survey established the extent to which scientific and technical reading involved federal employees or persons funded by the federal government. In that survey, we also determined whether page charges and journal subscriptions were paid through federal government grants and contracts, and we estimated the extent of use of numeric and bibliographic data bases. We also felt that there may be substantial federal expenditures involved in scientific and technical communication in the form of travel to and attendance at conferences, purchases of literature or information services, society memberships, and so on. In order to establish the extent of these expenditures, we sampled projects from the Smithsonian Science Information Exchange and surveyed the project directors to obtain this information. Other data are obtained from our previous studies.[26,27,28]

Table 8.12 Estimates of federal scientific and technical communication resources expenditures, by type of medium, 1977 (in millions).

Medium		Participants					
	Authors	Publishers	Libraries and A&I	User	Numeric Data Bases	Other	Total
Journals	$354	$21					
Other periodicals	281[a]	49					
Books	65						
Technical reports	294	13					
Other	191	92					
Total	$1,185	$175	$1,208	$2,296	$1,000	$502	$6,366

[a]Based on 1975 rates of 0.795 nonjournals per journal and assuming costs per periodical are equal.

Extent of Federal Expenditures in Scientific and Technical Communication

The extent of federal expenditures in scientific and technical communication is summarized in Table 8.12. From the author survey we established that 15 percent of the authors are employed by the federal government and these authors involve about $103 million in expenditures. Another 36 percent of the scientific and technical journal authors indicated that the research they reported was fully or partially funded by the federal government. We assume that their efforts come from these funds, which adds another $251 million to authorship expenditures. We also estimate the proportionate amount of authorship expenditures of other periodicals to be $281 million. From the survey of technical report authors, we estimate their expenditures to be about $294 million. The amount of federal expenditures for authorship of books and other forms of literature is assumed to be at the same proportion to the total as found with journals.

Three percent of journals and 10 percent of other periodicals are published by the federal government. Applying these proportions to the total publishing, we arrive at $21 million and $49 million for journals and other periodicals, respectively. We have no data on the federal involvement in book publishing. The publishing expenditures for federal-government-sponsored technical reports are taken from the National Technical Information Service budget and the proportion of the Government Printing Office budget that involves scientific and technical reports and monographs. Patent publications costs make up the other publication expenditures. The expenditures involving libraries and abstracting and indexing services consist of federal scientific and technical library expenditures estimates of $193 million, $82 million in direct federal subsidies to libraries an estimated $45 million expenditures involving abstracting and indexing services, and $768 million estimated to involve federally supported search services and $120 million for other library services. Estimates of the use expenditures involving federal funds come to $2,296. Estimates from a survey of numeric data bases give the federal expenditures for these data bases at $1 billion.

Estimates were made of the amount of other federal expenditures made in scientific and technical communication from such activities as attending scientific and technical meetings, membership in societies, and the like. About 2 percent of federal R&D funding is charged to the following scientific and technical communication categories:

1. Purchase of books, journals, articles, reports, bibliographies, and other text-type items on paper or microform.
2. Travel to and attendance at conferences, workshops, and seminars (travel, per diem, registration and other fees and charges).
3. Page charges and other fees and charges for professional, scientific, and technical journals.
4. Staff time and all service charges for on-line literature and/or data searches.
5. Purchase, rental, and lease of computer programs or data bases (bibliographic or numeric) on paper, microform, card, or tape.
6. Costs associated with printing and disseminating interim and final research reports.
7. Costs associated with producing and disseminating other information products (slides, tapes, film and so on).
8. Society memberships.
9. Salaries of librarians and other information specialists responsible for handling scientific and technical information.

As federal obligation for research and development in 1977 amounted to $23 billion, the total for these categories is estimated at $502 million.

The total for the categories above is $6.4 billion in 1977. Not included in this amount are estimates of professional salaries attributed to time spent in categories other than (4) and (9) in the list above or involved with the authorship of books, or periodicals other than journals. These costs would add another $475 million. Thus a figure of $6.4 billion federal contribution to 1977 scientific and technical

Table 8.13 Matrix of the flow of federal funds among journal system participants, 1977 (in millions).

	Govt. R&D	Govt. Authors	Other Authors	Govt. Publishers	Other Publishers	Govt. Libraries	Other Libraries	Govt. A&I	Other A&I	Govt. Readers	Other Readers	Total Payment
Government R&D	0	$107	$274	$21	0	$56	$9	$45	0	$580	$288	$1,380
Government authors	0	0	0	0	3	0	0	0	0	0	0	3
Other authors	0	0	0	0	20	0	0	0	0	0	0	20
Government publishers	0	0	0	0	0	0	0	0	0	0	0	0
Other publishers	0	0	0	0	0	0	0	0	0	0	0	
Government libraries	0	0	0	0.5	15	0.3	1	0	4	0	0	21
Other libraries	0	0	0	0.2	2	1	2	0	1	0	0	6
Government A&I	0	0	0	0	0	0	0	0	0	0	0	
Other A&I	0	0	0	0	0	0	0	0	0	0	0	
Government readers	0	0	0	0	0	0	0	0	0	0	0	
Other readers	0	0	0	0.3	2.3	0	0	0	0	0	0	3
Total Receipts	0	107	274	22	40	57	12	45	5	580	288	

communication appears conservative. It represents a 39 percent increase over our previous estimate of these federal expenditures in 1975. Even though the federal obligation for research and development increased more than 25 percent over the two-year period, a 39 percent increase in scientific and technical communication support is considered to be conservative, considering that the current estimate contains more elements than were included in 1975.

Flow of Federal Funds Among Journal Communication Participants

Since the amount of federal expenditures is so great, we felt it would be important to gain some insights concerning the flow of federal funds among communication participants. In order to do this, we concentrated on funds related to the journal system since we have better data in this area. This analysis is done by establishing the flow of federal funds between each of the major participants with each participant group subdivided by government and other participants (for example, authors employed by the federal government and those employed elsewhere but funded by the federal government). The flow of federal funds is depicted in Table 8.13, which shows the principal participants.

The cell entries are the amount of funds exchanged from the row participant to the column participant. For example, government authors are estimated to pay nonfederal publishers $3 million in page charges. The row totals, then, are the sum of payments of federal funds by the row participants. For example, the federal government libraries paid $21 million to other participants for journal products or services. The column totals are the receipts of federal funds by the column participants. For example, other (nonfederal) publishers received $40 million in federal funds from one participant group or another.

Our estimates suggest that the total amount of federal funds directly allocated to the participants is $1.4 billion. These are allocated to authors (or their organizations), federal publishers, libraries, abstracting and indexing services, and users.

We estimate from our author survey that 15 percent of the authors are government employees. Add to this the amount of funds they indicate that their employers spend on page charges, and we arrive at $107 million total. Similarly authors reporting on federally funded research yield an estimated $274 million. Based on the proportion of federally published scientific and technical journals, we arrive at an estimate of $21 million spent on publishing journals. The proportion of federal library budgets allocated to scientific and technical journals yields about $56 million, and subsidies to libraries is about $9 million from Higher Education General Information Survey (HEGIS). The amount of federal abstracting and indexing is estimated to be $45 million. The federal government readers and the federally funded readers comes to $580 million and $288 million, respectively, when individual subscriptions paid by the employers are added, for a rough total of $1.4 billion.

Authors employed by the federal government are estimated to pay $3 million in page charges to nonfederal publishers, and federally funded authors are estimated to pay $20 million. There is indication that a few other publishers also pay authors, but this amount is probably low. However, the copyright law could result in more being paid in royalties in the future.

We believe that federal government libraries pay about $500,000 to government publishers and $15 million to other publishers in subscription payments from federal funds. Federal government libraries also pay other libraries for photocopies made for interlibrary loans. This is estimated to be about $300,000 for government libraries and $1 million to other libraries. They also pay abstracting and indexing services for subscriptions at an estimated amount of $4 million. Other libraries are estimated to pay about $200,000 to federal publishers and $2 million to other publishers in subscription payments. However, these amounts could be substantially greater if indirect (or administrative) funding is included. It is only speculative as to how much of the institutional subscriptions are paid out of federal funds, but we think it could be as high as $200 million based on the proportion of federally provided research and development funding. Abstracting and indexing services sometimes pay publishers for subscriptions, but we do not know how much in federal funds might be involved.

Based on our reader survey, federal government employees pay a very small amount in subscriptions that are subsidized by their employers. On the other hand, federally funded researchers indicate that about $2.6 million is paid for subscriptions under federal grant or contract. We estimate that about $2.3 million of this goes to nonfederal publishers.

There are two major implications of the flow pattern of federal funds. First, the amount of the federal funds that ultimately go to other (nonfederal) publishers is not very great compared to the entire publishing costs. The total scientific and technical journal expenditures is estimated to be about $600 million, so the estimated $40 million received by them directly or indirectly from federal funds is only 7 percent. On the other hand, we were unable to identify the exact amount of indirect funds that are used for institutional subscriptions. We feel that this could be as high as $200 million if the share of institutional subscriptions is the same as the research and development funding. The same could be said about individual subscriptions since a substantial amount of scientific and technical employment is attributable to federal research and development funding.

The average flow of federal funds in scientific and technical communication presents a distorted picture in the sense that some publishers rely very heavily on page charges that are paid directly or indirectly through federal funds. These publishers might not survive otherwise. This presents another issue entirely. On the one hand, one might argue that a scholarly publication should not become reliant on such sources because of the uncertainty associated with it. Yet by using such a source of income, the publisher should reduce prices, thereby creating some additional subscriptions, which should yield some social benefit.

Results of the Federally Funded Projects Survey

The survey of federally funded research and development projects supplies data showing that each project produced an average of 1.98 articles in U.S. journals. If this number is expanded by the number of projects contained in the Smithsonian Science Information Exchange (from which the survey sample was drawn), the total number of articles generated by all government-funded projects is 158,400 articles published in U.S. journals. Almost 75 percent of respondents indicated that they planned to submit additional articles based upon the year's work. Assuming that the ratio of publishing in U.S. journals (U.S. in foreign journals) is the same for these plans as for articles actually submitted (84 percent), total U.S. articles published is increased to 208,000. At 158,000, federally funded projects would be accounting for 42 percent of all articles published in U.S. journals; at 218,000, for 5.5 percent. The estimate derived from the author survey (51 percent) falls within this range.

Additional information products for which number of items can be estimated from the survey show that in addition to articles in U.S. journals, an average research and development project produced 0.38 articles in foreign journals, 1.56 research reports submitted to the funding agency, 1.11 other written research products (such as books, monographs, and nonjournal periodical articles), 3.03 oral presentations in the United States, 0.75 oral presentations outside the United States, and 0.12 filed patents.

This pattern of production of scientific and technical communication products varies considerably by sector of performing organization and by field of science (Table 8.14). Academic performers

Table 8.14 Number of projects responding for number of specific information products generated, by field of science and performer, 1977.

	Articles in U.S. Journals		Articles in Foreign Journals		Other Written Research Products[b]		U.S. Oral Presentations		Foreign Presentations		Patents Filed	
	No.[a]	Mean	No.[a]	Mean	No.[a]	Mean	No.[a]	Mean	No.[a]	Mean	No.[a]	Mean
Total	660	1.98	481	0.38	641	1.11	676	3.03	508	0.75	674	0.12
Field of Science												
Physical Sciences	86	2.52	58	0.83	76	1.12	86	2.67	66	1.02	87	0.31
Mathematics	67	2.39	56	0.82	66	0.89	70	3.66	56	1.16	72	0.01
Computer Sciences	66	1.76	50	0.26	65	0.98	69	2.99	52	0.62	66	0.32
Environmental Sciences	84	1.94	62	0.34	85	1.78	83	2.80	63	0.71	85	0.04
Engineering	71	1.17	58	0.10	70	0.87	77	2.29	66	0.35	74	0.19
Life Sciences	74	3.01	49	0.51	73	0.97	76	2.91	53	0.91	76	0.03
Psychology	70	1.11	45	0.24	69	0.97	72	2.21	54	0.80	73	0.00
Social Sciences	77	1.27	62	0.11	80	1.36	78	4.15	53	0.45	79	0.15
Other Sciences	65	2.55	41	0.10	57	0.74	65	3.78	45	0.80	62	0.06
Performer												
Federal	200	1.62	132	0.17	190	1.47	207	3.03	158	0.59	206	0.18
Academic	281	2.59	208	0.65	269	1.11	282	3.31	205	1.02	285	0.07
Other	179	1.41	141	0.16	182	0.73	187	2.62	145	0.55	183	0.15

Source: King Research, Inc., Federally Funded Project Survey, 1978.

[a]Number of projects responding zero or more; excludes unknown number and no response.

[b]Excludes report submitted to funding agency.

tend to produce more of all types of products (excluding reports to funding agencies that have not been analyzed by field and performer) except for other written research reports, and patents. Federal performers lead in the production of both of these two types of products.

For articles in U.S. journals, Life Sciences projects produce the greatest number of articles, almost three times as many as do Psychology projects. Other leaders in production of U.S. articles are Other Sciences, Physical Sciences, and Mathematics projects. The pattern for producing articles in foreign journals is similar, except for the Other Sciences. In the category of Other Written Research Reports, Physical Sciences, Environmental Sciences, and Social Sciences are the leaders. Mathematics projects produce more oral presentations (combined foreign and U.S.) than do projects in other fields. Social Sciences projects provide the greatest number of U.S. presentations per project. Physical Sciences and Computer Sciences stand out as the leading producers of patents.

Additional research information products generated by federally funded projects include 17 percent of projects producing machine-readable files and 9 percent audio or visual media products. If one considers only one of these products each (since the actual number was not specified), there would be 13,600 machine-readable files (other than those for internal use only) and 7,200 audio or visual products from the 80,000 projects in the Smithsonian Science Information Exchange (SSIE) files.

REFERENCES

1. D. W. King, "A Potential Pitfall in the Economics of Information Products and Services," *Bulletin of the American Society for Information Science* 3(5):40 (1977).
2. F. Machlup, "Uses, Value, and Benefits of Knowledge," *Knowledge: Creation, Diffusion, Utilization* 1(1):62–81 (1979).
3. N. K. Roderer, "The Interchange of Scientific and Technical Information Between the U.S. and Other Countries" (Paper presented at the 41st ASIS Annual Meeting, New York, New York, November 13–17, 1978).
4. V. E. Palmour, M. C. Bellassai, and R. R. V. Wiederkehr, *Costs of Owning, Borrowing, and Disposing of Periodical Publications* (Arlington, Virginia: Public Research Institute, 1977).
5. S. V. Berg, "An Economic Analysis of the Demand for Scientific

Journals," *Journal of the American Society for Information Science* 23(1)23–29 (1973).

6. J. H. Kuney and W. H. Weisgerber, "Systems Requirements for Primary Information Systems: Utilization of *The Journal of Organic Chemistry*," *Journal of Chemical Documentation* 10(3)150–157 (1970).

7. J. A. Ordover and W. J. Baumol, "Public Good Properties in Reality: The Case of Scientific Journals" (New York: New York University and New York and Princeton Universities, 1975).

8. D. W. King, R. R. V. Wiederkehr, S. E. Bjorge, and P. F. Urbach, "Experimentation, Modeling, and Analysis to Establish a New Pricing Policy at CFSTI," *Proceedings of the American Society for Information Science* 5:311–314 (1968).

9. Charles River Associates, "Development of a Model of the Demand for Scientific and Technical Information Services" (Cambridge, Massachusetts: Charles River Associates, 1978).

10. D. D. McDonald, *Interactions Between Scientists and the Journal Publishing Process* (Rockville, Maryland: King Research, 1978).

11. Ordover and Baumol, "Public Good Properties."

12. J. H. Kuney, personal communication.

13. Ordover and Baumol, "Public Good Properties."

14. R. DeGennaro, "Escalating Journal Prices: Time to Fight Back," *American Libraries,* February 1977:69–74.

15. Machlup, "Uses, Value, and Benefits."

16. Berg, "Economic Analysis."

17. C. Herring, "A Study of Primary Journal Economics," in *Report of the Task Group on the Economics of Primary Publication* (Washington, D.C.: Committee on Scientific and Technical Communication, National Academy of Sciences, 1970).

18. Machlup, "Uses, Value, and Benefits."

19. W. J. Baumol and Y. M. Braunstein, *Scale Economics Production Complementarity and Information Distribution by Scientific Journals* (New York: New York University, 1976).

20. Ibid.

21. D. W. King and J. G. Yates, *Editorial Processing Centers: A Study to Determine Economic and Technical Feasibility: Annex Part V: Economic Analysis of Journal Publishing in the Life Sciences* (Rockville, Maryland: Westat and Aspen Systems Corporation, 1974).

22. Y. M. Braunstein, *An Economic Rationale for Page and Submission Charges by Academic Journals* (New York: New York University, 1976).

23. Capital Systems Group, *Page-Charge Policies and Practices in Scientific and Technical Publishing: A Historical Summary and*

Annotated Bibliography (Rockville, Maryland: Capital Systems Group, 1976).
24. Berg, "Economic Analysis."
25. Y. M. Braunstein, "An Economic Rationale for Page and Submission Charges by Academic Journals," *Journal of the American Society for Information Science* 28(6):355–358 (1977).
26. B. L. Wood, *Review of Scientific and Technical Numeric Data Base Activities* (Rockville, Maryland: King Research, 1977).
27. B. L. Wood and C. G. Schueller, *Results of the Federal Scientific and Technical Information Survey* (Rockville, Maryland: King Research, 1977).
28. D. W. King and D. D. McDonald, *Federal and Non-Federal Relationships in Providing Scientific and Technical Information: Policies, Arrangements, Flow of Funds and User Charges* (Rockville, Maryland: King Research, 1980).

Chapter 9

Hypothetical Economic Analysis of the Journal System

This chapter deals with decisions made by individuals and libraries concerning whether to purchase scientific and technical journals. Individuals can obtain needed articles from their own subscriptions, libraries, or reprints. The decision concerning which alternative to use depends partially on the cost of each. Similarly libraries can purchase journal subscriptions or borrow copies of needed articles, a decision that also depends partially on cost. Hypothetical examples are given concerning the cost trade-off for these two decisions. Borrowing copies (photocopies) by libraries involves copyright and royalty payments, which are also analyzed under hypothetical conditions. In particular, the effect of increased interlibrary lending and royalty payments is discussed.

AN INDIVIDUAL'S SUBSCRIPTION DECISION

Many factors enter into a scientist's or engineer's decision to purchase a journal subscription or to acquire articles through some other means. These include whether the subscription is part of a society membership, in which case the value derived is bundled with other benefits.[1] Other journal attributes enter this decision as well, including closeness of coverage to one's field of interest, price, editorial and review policies, graphics quality, and so on. These factors were studied by McDonald,[2] and Charles River Associates.[3]

McDonald examined three sets of variables with respect to individual scientists' decisions to subscribe to a journal: characteristics of the journals (such as circulation, price, and size), subscribers' past experience with the journals (such as number of articles read and number of articles submitted), and subjectively scored quantitative and qualitative journal attributes (such as benefit to career and emphasis on methodology). *Astrophysical Journal* subscribers identified twenty-seven other journals similar to it and these were separated into two categories: journals to which respondents subscribed and journals to which they did not subscribe.

The average numbers of articles submitted to subscribed versus nonsubscribed journals were not appreciably or statistically different. However, when asked what was the most they would be willing to pay for these other journals, respondents reported that they would be willing to pay approximately twice as much for their current journal subscriptions compared with the nonsubscribed journals ($43 versus $24). They also reported that they read slightly more articles in subscribed journals as opposed to nonsubscribed journals (twelve versus eight articles in a six-month period). So although it appears that subscribed journals were used more heavily than nonsubscribed journals (which the astrophysicists got primarily from library subscription), the difference in use was not large.

Nonsubscribed journals were generally larger (in terms of articles published per year) than subscribed journals, and they had slightly longer articles. Both sets of journals had approximately the same publication speed, suggesting that for these astrophysicists this variable did not contribute to a past subscription decision. There was some evidence that total circulations for the subscribed journals were higher than for the nonsubscribed journals, and additional evidence strongly suggested that individual subscription prices for nonsubscribed journals were actually higher than for subscribed journals. This again was evidence that small journals are generally more expensive.

When respondent-scored journal attributes were compared between subscribed and nonsubscribed journals, the following five were identified as the most important for *Astrophysical Journal* subscribers, both in terms of their overall importance to the subscriber and in terms of their ability to differentiate between subscribed and nonsubscribed journals:

1. The journal regularly publishes papers by a large proportion of all the people about whose work I want to know.
2. The price for a one-year personal subscription to the journal is low.
3. The journal seldom publishes work in my field that is obsolete or out of date.
4. The journal places a strong emphasis on theory.
5. The average length of papers published by the journal is low.

The three least important journal attributes for astrophysicists were:

1. The average length of papers published by the journal is low.
2. The journal places a strong emphasis on methodology.
3. The journal accepts a large proportion of all papers submitted to it.

Combining the first set of five attributes in an additive multiattribute utility model (see Chapter 3) suggested that the utility of the other subscribed journals was not statistically different from nonsubscribed journals, even though *Astrophysical Journal* itself received higher utility scores than other subscribed journals. McDonald concluded that the utility modeling approach for describing individual subscribers' decisions was promising but might be significantly improved by incorporating both probability assessments and attributes directly related to journal subject content.

McDonald's is an example of a multivariate approach to predicting individual scientists' subscription decisions. Here we will look at other facets of the issue dealing with the economic trade-off of purchasing a journal subscription versus using another source because this issue provides some insights about the economics of individual decisions and the overall journal system.

For simplicity, we will analyze the cost of individual use in the absence of other factors that enter into one's decision to purchase a journal subscription. We assume that a scientist or engineer can obtain a copy of an article through a personal subscription, from a library, or from an author reprint. For the most part it appears to be more desirable to use one's own subscription, but there are instances in which the needed article is in a journal that has fewer other articles of interest or for which the scientist or engineer lacks a personal copy because it is lost or discarded or because the article was published prior to subscribing to the journal. In the latter instance, the scientist or engineer can obtain a copy from a library or author (assuming a colleague does not have a copy).

If the scientist or engineer obtains a copy from a library, the cost elements consist of time for searching, for going to a library and obtaining the copy, for photocopying the article, and for reading the article. If the article is obtained from the author, the cost includes time for searching, locating the author, writing and mailing a request, and reading the article. Here, however, a penalty is paid in terms of time delays. The cost of reading one's own subscription includes the subscription price, acquisition and storage, searching, and reading time.

Table 9.1 Average cost per reading of scientific and technical journals, 1977.

No. of Readings	Total Cost	Cost per Reading
1	$43.80	$43.80
2	55.30	26.80
3	66.40	22.10
4	77.50	19.40
5	88.60	17.70
6	99.60	16.60
7	110.70	15.80
8	121.80	15.20
9	132.90	14.80
10	143.00	14.40
11	155.00	14.10
12	166.10	13.80
13	177.20	13.60
14	188.30	13.40
15	199.40	13.30

If one assumes that a scientist takes about fifteen minutes to obtain a copy in the library and, say, five minutes to order an article reprint, the total direct cost to a scientist or engineer per reading comes to $14.50 for a library use and $10.80 for a reprint use. The equivalent cost per reading of an individual journal subscription depends on the number of readings by an individual of that subscription. This cost per reading is calculated in Table 9.1. These costs reflect only those incurred by the scientist, mostly through time expended. Analysis on a total system basis would substantially increase the cost of using a library copy and also modify the other two categories of cost somewhat. If we compare the cost per reading by different paths, it is found that it is less expensive for an individual to purchase a subscription than to use a library if there are more than ten readings from the journal subscription. With fewer than that number, it is economically best to rely on the library. This is not too different from what may be happening in actuality. On the average, there are 11.3 readings per individual journal subscription, with the remaining readings coming from library subscriptions or reprints. With reprints, it appears that there are further considerations; such as the time delays and uncertainty involved in obtaining reprints and the added benefits derived from the total journal issue rather than an individual article.

Since the primary fixed cost is the journal subscription price, we will look at the effect of the difference in journal prices of large journals and small journals. The average price of a small subscription journal is

Table 9.2 Individual's break-even point for purchasing subscription or using a library or reprint, by size of journal, 1977.

Size of Journal	Purchase vs. Library Break-even Point (No. of Uses)	Purchase vs. Reprint Break-even Point (No. of Uses)
Small	12	12
Medium	9	9
Large	6	6

$41.62, it is $31.83 for medium journals, and $21.47 for large journals. The break-even point is shown in Table 9.2. According to the table, a scientist or engineer must have twice as many readings in a small journal as in a large journal to make it worthwhile to purchase a subscription. This may partially explain why the proportion of domestic subscriptions to libraries is much higher for small journals than large journals (59 percent small, 34 percent medium, 29 percent large). The highest priced journals by far are found in the Physical Sciences ($96.83). They also have the lowest average number of individual subscriptions per scientist (1.2 subscriptions per scientist compared to 5.9 for all fields) and the lowest proportion of individual subscriptions (31 percent compared to 56 percent for all fields).

This simple analysis presents a paradox. Ideally we would like a journal system that always distributes needed articles directly to scientists and engineers. The journal system that packages articles in bundles and sends them to scientists is fairly efficient. However, some unwanted articles are distributed by this means. Fortunately there is a backup system in libraries and reprints that can be employed by scientists and engineers. But even here, economics suggests that libraries not acquire all journals read by their patrons but rather sometimes rely on interlibrary loan.

A LIBRARY'S SUBSCRIPTION DECISION

One current trend is an increase in library networking and resource sharing involving interlibrary loans of journal articles. Many libraries are finding that it is less expensive to borrow copies of journal articles from rarely used journals than to subscribe to these journals. This section gives some implications of this trend based on a model that determines the trade-off of using interlibrary loans versus purchasing journal subscriptions. The bases for this model are the distribution of use of

Table 9.3 Number of journals and total readings, by number of
readings, 1977.

No. of Readings	No. of Journals	Total No. of Readings
0	126	0
1	65	65
2	32	64
3	22	66
4	18	72
5	14	70
6	7	42
7	7	49
8	7	56
9	7	63
10	4	40
Over 10	51	997

journals in libraries and the comparative unit costs of in-house use and
interlibrary loans. Data for the distribution of use are derived from a
study performed at the University of Pittsburgh.[4] For the purpose of
analysis, we assume a typical library subscribes to about 360 scientific
and technical journals. The average readings per journal volume in a
subscribing library is 4.4. The Pittsburgh data were adjusted to achieve
this average (Table 9.3).

It is assumed that interlibrary loans could approach an optimum
system (from the standpoint of libraries) in which costs are minimized.
This can be achieved by not initially purchasing journals that have little
potential for use and by canceling existing subscriptions to journals
when it is found to be less expensive to borrow them.

We will work with data from studies by Williams et al.[5] and Palmour
et al.,[6,7] which were derived to determine at what point it becomes more
practical for libraries to borrow rather than buy journals. The Williams et
al. cost data are somewhat dated since that study was done in 1968, so
we made adjustments for inflationary increases. The adjusted data
correspond closely to those used in Palmour et al. Costs for 1977 are
given in Table 9.4. It is assumed that journal storage costs will occur
over twenty years and that the present value of initial journal costs can
be compared over twenty years using a 5 percent compound interest
rate.

Accumulating costs over journals comes to $70.51 per journal and
accumulated costs of use come to an additional $11.60 per reading. The
cost of in-house use and interlibrary loan is given in Table 9.5. The cost
per reading of purchased copies is higher than interlibrary loan below
five readings (when both the cost of borrowing and lending is included).

Table 9.4   Costs of library operations associated with journals, 1977.

Processing	
Acquisitions	$95.91 per new journal
Price of journal	$32.69
Annual maintenance (check-in, binding, file maintenance, etc.)	$30.92 per journal (excluding subscription price)
Storage	$6.00 per journal
Weeding	$0.90 per journal
Use	
Internal use and circulation	$2.00 per use
Reading	$9.60 per reading
Interlibrary loan	$11.60 (borrowing library) + $8.40 (lending library) = $20.00

Thus in the absence of other factors such as delays in interlibrary lending and reduced capability to browse, below that level it would be less expensive to borrow copies than to renew a subscription. It is noted that an estimated 73 percent of the journals have fewer than five readings and this represents only 17 percent of the total readings.

An analysis is given below to illustrate the effect of libraries' incorporating a policy of continuing journal subscriptions only when they have five or more readings. This analysis includes costs for prerun, runoff, acquisition, storage, circulation and interlibrary loans, and reading for all scientific and technical journals. The average number of subscriptions is estimated to be 6,327 per journal, which includes 1,789 library subscriptions, 3,542 individual subscriptions, and 996 foreign subscriptions. The total prerun cost is estimated to be $324 million, and the runoff cost is $202 million over all journals. If we subtract the portion of income allocated to sources other than subscriptions* and the portion attributable to other subscribers, we arrive at $92 million and $57 million for prerun and runoff, respectively. Thus total publishing cost allocated to library subscriptions is $149 million (Table 9.6). Maintenance cost of 1,789 library subscriptions over all journals comes to $246 million, storage costs are $48 million, and internal use cost comes to $66 million and $385 million with user reading costs included. There are currently estimated to be 2.9 million interlibrary loans[8] of scientific and technical journals. These loans yield costs of $24 million

---

*Income from subscriptions is only about 82 percent of the total, with other sources coming from page charges; sale of back issues, reprints, and microform; and advertising. The values are subtracted from the total since the costs are not borne by the libraries.

Table 9.5 Cost of purchasing journals and interlibrary loans over a number of uses, 1977.

No. of Readings	No. of Journals	Total No. of Readings	Journal Cost	Reading Cost	Cost per Reading	Journal Cost	Interlibrary Loan	Cost per Reading	Savings
0	126	0	$8,900	$0	$0	$0	$0	$0	$8,900
1	65	65	4,570	750	81.80	0	1,920	29.60	3,400
2	32	64	2,260	740	46.90	0	1,890	29.60	1,100
3	22	66	1,550	770	35.70	0	1,950	29.60	370
4	18	72	1,270	840	29.30	0	2,130	29.60	−20
5	14	70	990	810	25.70	0	2,070	29.60	−270
6	7	42	490	980	23.30	0	1,240	29.60	−230
7	7	49	490	570	21.60	0	1,450	29.60	−390
8	7	56	490	650	20.40	0	1,660	29.60	−520
9	7	63	490	730	19.40	0	1,860	29.60	−640
10	4	40	280	460	18.60	0	1,180	29.60	−440
Over 10	51	997	3,600	11,570	15.20	0	29,510	29.60	−14,340

Table 9.6 Current costs of circulation and interlibrary loan for scientific and technical journals, 1977 (in millions).

Participant	Prerun Costs[a]	Runoff Costs[a]	Maintenance Costs[b]	Storage Costs	Internal Use Costs	Interlibrary Loan Costs	Total Costs
Publisher	$92	$57					$149
Lending library						$24	24
Borrowing or using library			$246	$48	$385	62	741
Total	92	57	246	48	385	86	914

[a]Includes profit and other sources of income subtracted in proportion.
[b]Does not include subscription price.

Table 9.7 Total costs of circulation with optimum interlibrary loan for scientific and technical journals, 1977 (in millions).

Participant	Prerun Costs	Runoff Costs	Maintenance Costs	Storage Costs	Internal Use Costs	Interlibrary Loan Costs	Total Costs
Publisher	$92	$17					$109
Lending library						$76	76
Borrowing library			$66	$13	$314	191	584
Total	92	17	66	13	314	267	769

to the lending libraries and $34 million plus $28 million for reading to the borrowing libraries. Thus the total costs to libraries comes to $914 million or $25 per use over the 36.1 million readings from library journals.

If all libraries purchased only journals with five or more uses, there would be about 9 million interlibrary loans and 27.1 million internal uses (see Table 9.7). The number of journal subscriptions would be reduced by about 73 percent. Although the prerun costs would remain the same, runoff costs would go down since the number of library subscriptions would be reduced. There would now be 5,021 total subscriptions instead of 6,327. The runoff costs would be $17 million. Since fixed costs are now allocated over a smaller number, the price of subscriptions would increase. If the drop in income is allocated to all subscriptions, there would be a 16 percent increase in price across all subscriptions. If the drop is allocated only to library subscriptions, the increase in price would need to be about 170 percent. The economic effect of the price increase on demand is ignored for this analysis. Maintenance costs are reduced to $66 million, storage costs come down to $13 million, internal use costs are now $314 million, and interlibrary loan costs become $267 million. Total costs are now $769 million, or $21 per reading.

Under the assumptions made, the overall effect of an optimum interlibrary loan network system would be a reduction in cost to the library per reading of about $4 (16 percent). In effect, through interlibrary loans, libraries will reduce their in-house costs by $145 million, including initial subscription prices and costs to the lending libraries. However, lending libraries are beginning to charge for their services. Also, ultimately, prices of the journal would have to increase some to recover publishing costs. Copyright royalty payments may provide some of the income to publishers that would recover some of the lost subscription revenue. Therefore gains may be only temporary.

There is quite a difference in analysis results when one looks individually at the results for small, medium, and large journals. First, the price of small journals is much higher than large journals. Thus the break-even point of number of uses goes up for small journals, so, other things being equal, a higher proportion of them will have their subscriptions canceled. Second, small journals are much more dependent than large journals on library subscriptions since a higher proportion of their subscriptions come from libraries (institutions). This again means more cancellations, relatively speaking, for small journals. The break-even point of number of uses for the three sizes of journals is given in Table 9.8. The break-even point is greater than five, four, and four uses for the small, medium, and large journals, respectively. Assuming an optimum strategy, the proportion of journal subscription

Table 9.8 Average cost per reading with library purchase of journals, 1977.

Number of Uses	Size of Journal		
	Small	Medium	Large
0			
1	$91.00	$81.25	$70.90
2	51.30	46.42	41.24
3	38.10	38.82	31.40
4	31.52	29.01	26.40
5	27.50	25.53	23.46
6	24.84	23.21	21.48
7	22.95	21.55	20.07
8	21.53	20.31	19.01
9	20.43	19.34	18.19
10	19.54	18.56	17.53

cancellations would be 45, 27, and 26 percent, respectively. These results applied to all scientific and technical journals are displayed in Table 9.9. It is clear that the loss in income to small journals would be substantially greater than to medium or large journals if libraries were to rely on interlibrary loans to the optimum extent (always borrowing when it is less expensive than to purchase journals). This picture is probably even worse since small journals have relatively less income from other sources, such as page charges, advertising, and sales of back issues. With higher prices, the break-even point for small journals will go up even further, thereby requiring more canceled subscriptions. The copyright law (and photocopying royalty payments) is thought to help this situation, but it can actually act to the disadvantage of small journals.

EFFECT OF THE NEW COPYRIGHT LAW ON SCIENTIFIC AND TECHNICAL JOURNALS

One of the most significant recent events that could alter the distribution of scientific and technical information is the new copyright law, which went into effect in January 1978. It includes a section dealing with photocopying of library materials. In effect, this section can be interpreted to suggest that photocopying of journal articles may be subject to royalty payment to authors or publishers, except under certain eligibility conditions. For example, the national Commission on New Technological Uses of Copyrighted Work (CONTU) set forth

Table 9.9 Effect of interlibrary loans on small, medium, and large journals, 1977

Information	Size of Journals		
	Small	Medium	Large
Number of journals	2,271	1,580	596
Average number of subscriptions per title	1,391	5,454	27,450
Average number of institutional subscriptions per title	626	1,468	7,069
Proportion of subscriptions that are institutional	0.45	0.27	0.26
Average number of institutional subscriptions canceled	457	1,072	5,160
Proportion of subscriptions canceled	0.33	0.20	0.19

guidelines concerning photocopying for interlibrary loans. They suggest that the borrowing library be responsible under the law for photocopies made for their interlibrary loan requests. Conditions of eligibility are that such photocopies made for fair use and in any given year for fewer than six copies of a journal published fewer than six years ago are exempt from royalty payments. Fair use includes such photocopies as those made for classroom use by faculty and those made for replacement of owned materials. Photocopies made for intrasystem loans (to branches of a single library organization) or for local users may also be subject to royalty payment. However, there are yet no clear-cut interpretations or guidelines, such as the CONTU proposed, to determine conditions of royalty eligibility for photocopying for local users or for intrasystem loan. The amount of photocopying performed under various conditions of eligibility for all three types of transactions (interlibrary loan, intrasystem loan, and local use) is given in a King Research report.[9] These data indicate the amount of photocopying performed on scientific and technical journals. They show that the copyright law could have a severe effect on the journal publishing industry, particularly on small journals, which are less economically stable than large journals. Results are sketchy and require some broad assumptions. However, they do point out some potential areas of concern.

In the King Research report on library photocopying, the principal observation was the volume of photocopying of library materials performed by library staff. However, another part of the study gives some evidence concerning how much additional photocopying of journals might be performed on the same photocopying machines by users and on machines that are provided for exclusive use by library patrons. Finally the report also provided some evidence of the proportion of photocopying that was performed on scientific and technical periodicals (Table 9.10). There are three kinds of machines in libraries: those that are used only by library staff, those used by both library staff and patrons, and those used exclusively by patrons. The first column gives the amount of photocopying of all periodicals that is done only by library staff on the first two kinds of machines. The second column gives an estimate of the amount of additional photocopying of periodicals performed on the machines used by both library staff and patrons. These figures are based on the total exposures made on the first two categories of machines. The third column is the estimated total with photocopying on machines available only to patrons included. The fourth column is the proportion of photocopying that is from scientific and technical periodicals. The last column is the estimated total amount of photocopying performed of copyrighted, domestic, scientific and technical periodicals.

Table 9.10 Number of photocopies of periodicals in U.S. libraries, by type of library, 1976.

Type of Library	Domestic Copyrighted Periodicals (millions)	Total, including Patron Use on Machines (millions)	Total, including Patron Only Machines (millions)	Proportion S&T (%)	Total S&T Photocopying (millions)
Public	12.88	19.22	26.74	46	12.30
Academic	3.74	5.58	15.37	65	9.99
Special	12.82	25.64	35.46	87	30.85
Federal	2.52	5.04	8.83	89	7.86
All	31.96	55.48	86.40	71	61.00

Source: King Research, Inc., *Library Photocopying in the United States: with Implications for the Development of a Copyright Royalty Payment Mechanism* (Washington, D.C.: Government Printing Office, 1977), pp. 64,78,91,147.

Table 9.11   Number of photocopies of scientific and technical periodicals subject to royalty payment, by type of transaction, 1976 (in millions).

	Interlibrary Loan	Intrasystem Loan	Local Use	Total
Total photocopies	2.9	8.6	17.5	29.0
Royalty payment				
High estimate	1.8	7.7	15.8	25.3
Low estimate	0.5	2.9	4.4	7.8

From the user survey, there were estimated to be 29 million photocopies of articles from the 4,447 scientific and technical journals in 1977. Extrapolating this figure to the larger class of 8,915 scientific and technical periodicals, which includes both journals and other types of periodicals, the estimate is 58 million. This compares closely with the 61 million photocopies in 1976 estimated in Table 9.10.

The question then becomes what portion of the 29 million photocopies of journals is subject to royalty payment. Of these 29 million photocopies, 2.9 million are estimated to come from interlibrary loan, 8.6 from intrasystem loan, and 17.5 million from local use. By applying both loose and stringent interpretations of the copyright law and CONTU guidelines, we arrive at high and low estimates of number of photocopies subject to royalty payment (Table 9.11). We emphasize that these illustrations are hypothetical, since interpretations of what is eligible for payment are still under discussion at this writing.

Our estimates suggest that as many as 25 million or as few as 8 million photocopies may be subject to royalty payment. A study by Indiana University[10] indicates that publishers who desire royalty payment indicated on the average about $1.00 to $1.50 as the expected fee. If one adds the cost of maintaining records, the sum of costs and fees might be about $1.50 per royalty payment, for a total of a $12 million to $38 million cost related to royalty payments. If these costs are all borne by libraries, they may have several alternatives concerning where to obtain funds to cover these expenditures. Examples from the Indiana University[11] and King Research[12] studies suggest that possible actions might include cancellations of subscriptions (either multiple subscriptions to single journals or infrequently used journals), budget increases, or budget reallocations. The average library subscription price for scientific and technical journals is $37.72. If all royalty payment costs result in canceled subscriptions, the number could be as high as 1.5 million or as low as 520,000 cancellations. The total number of library subscriptions is estimated to be about 8 million so the potential number of subscription cancellations is not insignificant (6 to 19 percent). If uses of canceled subscriptions come from interlibrary loans, additional cancellations will be required to cover these costs.

We can make some conjectures from the little evidence we have as to the direction, if not magnitude, of changes resulting from the copyright law. Ideally income from journals should be distributed by their worth, or at least by their amount of use. The new copyright law takes a small step in the latter direction since use is partially determined by the amount of photocopying that takes place. However, the royalty payments inject additional funds into the economic system, and they are redistributed among current journals. Also it is possible that some photocopying will not be reported or that library staffs will discontinue or reduce the amount of photocopying that they do. They may switch to unmonitored or coin-operated machines to reduce their liability. Or library patrons may shift from using library subscriptions to purchasing their own. Undoubtedly a combination of these actions (as well as others) will take place. However, if a significant amount of royalty payments are made, we feel that there will be some redistribution among current journals.

Each individual journal will gain and lose some. Gains will be achieved through photocopying that is performed on internally owned issues. However, nearly every journal is subject to subscription cancellations in some libraries where it is infrequently used. It is assumed that uses for such cancellations will come from interlibrary loan. We have seen that increased interlibrary loans may be to the detriment of small journals. However, all sizes of journals will be affected to some extent if the current amount of interlibrary loans by size of journals is projected in proportion to the current mixture of loans. In Table 9.12, the average number of interlibrary loans is given for small, medium, and large journals. These loans are for the 132 libraries served by MINITEX.* Here the journals were classified as were the 4,309 scientific and technical journal population used in this report. Circulations for the MINITEX sample of journals were obtained from *Ulrich's*. A total of 93 journals (53 percent) had circulation given in *Ulrich's*. The average number of loans ranges from 9.1 to 10.0 for the three sizes of journals, suggesting that all three sizes have about the same level of interlibrary loan. However, since there are more small journals, the average number of loans per library is greater for the small journals.

*MINITEX is an organization in Minnesota that is responsible for enhancing and expediting interlibrary loans in Minnesota and surrounding states.

Table 9.12 Average number of scientific and technical periodical items borrowed, by size of journal, 1975–1976.

	Size of Journal		
	Small	Medium	Large
Average per title	9.3	10.0	9.1
Average per library	3.6	1.8	1.2

Table 9.13 Distribution of size of scientific and technical journals borrowed and journals in the population.

	Size of Journal			
Proportion of Titles	Small	Medium	Large	Total
All U.S. S&T journals	0.51	0.35	0.13	0.99
MINITEX journals	0.55	0.27	0.18	1.00

The question then is whether the journals in MINITEX are typical of the 4,447 scientific and technical journals published in the United States in 1977. The estimate of the number of these journals in MINITEX that have at least one loan is 5,000. (That number, however, includes some journals no longer in existence). The proportions of scientific and technical journals and MINITEX journals falling into the three sizes are displayed in Table 9.13. The distribution of size of U.S. and MINITEX journals is similar. Thus there is no evidence that MINITEX journals are atypical.

If there is a budget squeeze caused by royalty payments, some subscriptions will be canceled, and the number of interlibrary loans should go up for infrequently used journals. However, the CONTU guidelines suggest that no royalty payments be made on journals unless the borrowing library has over five photocopies made. Thus there is some chance that few or no royalty payments will be paid to journals for increased interlibrary loans. One exception to this might be if a central library facility such as the proposed National Periodicals Center[13,14] is ruled to have to pay royalty fees. In that case, small journals might make arrangements with such organizations in order to recover their losses.

Increased interlibrary lending with the cancellation of infrequently used library subscriptions could yield substantial savings to libraries (and to the entire journal system). However, publishers must increase their prices substantially to cope with the reduced number of subscriptions. If they do this across all subscriptions, the number of individual subscriptions and library subscriptions will go down slightly. However, in both instances the break-even points change only slightly. If, on the other hand, all of the subscription increases go to libraries, the break-even point increases to about seven, which potentially means another 8 percent of the journal subscriptions would be canceled. On the other hand, if interlibrary loans come from a central system and libraries pay a royalty fee, the prices would not need to be increased. First, we assume that the optimum economic conditions will occur and 9 million interlibrary loans will be made and that lost income due to canceled subscriptions will be recovered through royalty payments. Under these conditions, the necessary royalty payments would have to be $7.50. If this happens, the amount of interlibrary loan costs would increase from $20.00 to $27.50, and the break-even point would decrease by one use.

The total amount paid by libraries is about the same, regardless of the payment procedure used. We anticipate that the amount of borrowing will not reach the levels mentioned above and therefore the prices or royalty payments will not be as high as given above. However, if the borrowing versus purchasing decision is based on the use criterion, the optimum economic approach would be achieved from the standpoint of the entire journal system. In some instances, the amount of royalty payments could be high—in the $5.00 to $10.00 range—for some small journals in order to survive. But in the long run, it will be necessary to pay these royalty payments to achieve the best overall journal system.

REFERENCES

1. J. A. Ordover and W. J. Baumol, "Public Good Properties in Reality: The Case of Scientific Journals" (New York: New York University and New York and Princeton Universities, 1975).
2. D. D. McDonald, *Interactions Between Scientists and the Journal Publishing Process* (Rockville, Maryland: King Research, 1979).
3. Charles River Associates, *Development of a Model of the Demand for Scientific and Technical Information Services* (Cambridge, Massachusetts: Charles River Associates, 1978).
4. A. Kent, K. L. Montgomery, J. Cohen, J. G. Williams, S. Bulick, R. Flynn, W. N. Sabor, and J. R. Kern, "A Cost-Benefit Model of Some Critical Library Operations in Terms of Use of Materials" (Pittsburgh, Pennsylvania: University of Pittsburgh, 1978).
5. G. Williams, E. C. Bryant, R. R. V. Wiederkehr, V. E. Palmour, and C. J. Siehler, *Library Cost Models: Owning Versus Borrowing Serial Publications* (Rockville, Maryland: Westat Research, 1978).

6. V. E. Palmour, M. C. Bellassai, and R. R. V. Wiederkehr, *Costs of Owning, Borrowing, and Disposing of Periodical Publications* (Arlington, Virginia: Public Research Institute, 1977).
7. Task Force on a National Periodicals System, *Effective Access to the Periodical Literature: A National Program* (Washington, D.C.: National Commission on Libraries and Information Science, 1977).
8. King Research, *Library Photocopying in the United States: With Implications for the Development of a Copyright Royalty Payment Mechanism* (Washington, D.C.: Government Printing Office, 1977).
9. King Research, *Library Photocopying.*
10. B. M. Fry, H. S. White, and E. L. Johnson, "Scholarly and Research Journals: Survey of Publisher Practices and Present Attitudes on Authorized Journal Article Copying and Licensing" (Bloomington, Indiana: Graduate Library School, Indiana University, 1977).
11. Ibid.
12. King Research, *Library Photocopying.*
13. Palmour, Bellassai, and Wiederkehr, *Costs of Owning.*
14. Task Force on a National Periodicals System, *Effective Access.*

Chapter 10

The Future of the Scientific and Technical Journal System

From a technical and systems standpoint, the scientific and technical journal has not changed a great deal in the past three hundred years. The published journal is still similar to those that appeared in the 1600s. In the United States, the journal publishing industry has grown steadily over the past two decades and is anticipated to continue to do so. However, a number of current events could dramatically alter the journal system. First, a change is taking place concerning the quantity of articles that are distributed in separate copies. This includes both interlibrary loans and individual separates (preprints, reprints, and photocopies) distributed by authors and publishers. Advances in library networking and resource sharing may soon reach the point where there will be a formal national periodicals system in this country.[1,2] Exciting advancements in electronic processing taking place in such areas as office word processing, telecommunication, technology for nonscientific publication, on-line bibliographic searching, minicomputers, and mass digital storage will almost certainly yield enormous advantages to future scientific and technical communication. In this chapter we examine the potential future (to the year 2000) of the journal system in light of the rapidly changing environment.*

THE FUTURE OF AUTHORSHIP

Science and technology are expected to continue their steady growth through the end of this century. In the past five years there has been an upswing in U.S. research and development funding (annual growth rate of 8 percent) and in the number of U.S. scientists and engineers (2 percent). This growth, which will probably continue at least to 1985, is partially expected because of a current favorable attitude toward science and technology by the public, industry, and government. Also the gross national product is expected to increase by almost

*Much of this discussion is derived from a National Science Foundation study on the electronic alternative to communication through paper-based journals.[3]

Table 10.1 Projection of number of scientists and engineers and number of articles published, by field of science, 1975–1985.

Field of Science	No. of Scientists or Engineers (000)					No. of Articles Published (000)				
	1975	1977	1980	1985	Ten-Year Change (%)	1975	1977	1980	1985	Ten-Year Change (%)
Physical Sciences	275	282	285	290	5	56.0	57.5	58.2	60.8	9
Mathematics	108	116	131	156	44	8.1	8.6	9.3	10.0	23
Computer Sciences	171	196	242	315	84	8.2	9.0	9.7	11.2	37
Environmental Sciences	83	89	100	118	42	14.7	17.0	20.4	26.3	79
Engineering	1,340	1,383	1,402	1,424	6	43.8	46.6	49.7	56.9	30
Life Sciences	308	321	339	361	17	114.8	127.8	144.4	172.0	50
Psychology	119	132	160	208	25	12.4	13.4	15.7	19.5	57
Social Science	231	260	333	461	100	53.6	57.1	58.6	60.2	12
Other						42.1	45.2	50.6	54.9	40
All Fields	2,635	2,779	2,992	3,333	26	353.7	382.3	416.7	476.8	35

150 percent to $3.7 trillion by 1985 (an optimistic forecast that could be blunted by unforeseen events).

In the past, growth in science and technology has yielded a corresponding increase in communication activity, including journal production and use. Thus with the expansion in number of scientists and engineers, we expect a steady increase in the number of articles written and published up to 1985 (Table 10.1). This increase is greatest in Environmental Sciences (79 percent between 1975 and 1985) and least in Social Sciences (12 percent). This increase will also be reflected in the number of journals published, but not at quite the same rate.

Advances in technology that will have a significant effect on authorship are electronic word-processing and text-editing systems. (They will also have a second-order effect on other system functions.) These systems have already achieved widespread use. Word-processing systems use editing typewriters that use magnetic digital storage of one form or another. There are also text-editing systems that range from terminals that tie into a local computer to intelligent terminals with self-contained memory and microprocessor-based computing functions. While these systems help achieve improvements in editorial quality of manuscripts and provide faster and more reliable input, their most direct advantage comes from the substantial reduction of secretarial labor costs. The economics involve a cost trade-off between increased investment in equipment against reduced labor. Keen competition in this area is resulting in increased sophistication and flexibility of equipment as well as in their reduced costs. Options such as editing and spelling routines, special print stations, message service, and, perhaps most importantly, photocompositor interfaces are available. In the long run, significant numbers of manuscripts will be transmitted to publishers in digital form, and articles will be edited and composed electronically, substantially reducing prerun publishing costs.

It has been estimated that the general use of terminals will more than double from 1975 to 1980,[4] and by 1990 most handling of business records and correspondence will be electronic.[5,6] Since many organizations will obtain electronic systems for various uses, the capital investment for purchasing or leasing the equipment will be spread over a broad base and the cost of preparing journal article manuscripts may well be lower than by traditional typewriter. Also it is possible that library minicomputers, terminals, and other equipment will be used by authors, just as libraries now assist authors through on-line bibliographic search services.

Other advanced technologies that may assist authorship are teleconferencing and on-line bibliographic systems. On-line computer

Table 10.2 Projections of electronic processes used and cost of authorship, 1975–1985.

	1975	1977	1980	1985	Ten-Year Change (%)
Professional salaries (hourly)	$8.32	$9.60	$11.44	$14.52	74
Secretarial/clerical (hourly)	$4.52	$5.28	$6.38	$8.67	92
Support services Articles prepared by typewriter (%)	60	55	45	20	−67
Articles prepared by word processing (%)	25	30	35	50	100
Articles prepared by text editor (%)	15	17	20	30	100
Electronic savings per article	$45.00	$49.00	$56.00	$70.00	56
Author cost per article	$1,540	$1,780	$2,120	$2,720	77
Total authorship cost ($ Million)	$405	$534	$696	$1,017	151

systems are being used more frequently as a medium of interpersonal communication. This form of communication can upgrade authorship by facilitating much more extensive informal feedback, which in turn supports the preparation of a manuscript before submission to a publisher. The increase in preparing manuscripts in digital form will complement or enhance teleconferencing. On-line bibliographic searches probably should increase the quality of background research and, thus, the quality of articles. The advances in electronic technology will also reduce costs somewhat in the future. However, overall costs of authorship will continue to increase up to 1985 since the principal elements are professional and support labor. The projections of these cost rates are given in Table 10.2 along with estimates of average cost per article and total costs. We also indicate savings that are attributable to electronic processes.

In 1980 it is anticipated that the majority of authors will have word-processing or text-editing systems available to them, and about 55 percent of the articles will be prepared electronically. By 1985 this proportion will increase to about 80 percent. It is significant that most articles are written by authors affiliated with universities or colleges and that, according to a recent study,[7] academic institutions lag slightly behind other organizations in the use of electronic equipment such as terminals and word processing systems.* These factors suggest that the future use of electronic processes for authorship may not be quite as high as for general uses that are less concentrated in academic institutions.

We estimate that the use of electronic processes will result in some savings (Table 10.2). Overall the average cost of articles will increase about 77 percent, compared with a projected 74 percent increase in professional salaries and a 92 percent increase in support salaries. The savings are helpful but certainly do not dominate costs. The average cost per article, $1,540 in 1975, is estimated to increase to $2,720 in 1985. This is an increase of 77 percent in current dollars and 25 percent in constant dollars.

We expect that between 1985 and 2000, nearly all article manuscripts will be handled by word-processing and text-editing systems and thus could be sent to publishers in digital form. We predict that on-line bibliographic search and teleconferencing will be commonplace for scientists and engineers. The available electronic processes in authorship can also be expected to yield a greater degree of integration of research dissemination than was previously found in informal memoranda, conference papers, technical reports, journal articles, and books, which will substantially improve the quality of dissemination of research results.

Some major human-factor barriers are expected to persist. For example, author motivation is partially based on recognition and the pressure to publish. If and when publishers begin to distribute separates on demand, authors may become reluctant to submit their manuscripts to them. Logic implies that the importance of a published article lies not in the graphic or paper quality of the journal or in its circulation but rather in the quality of its content, which is established by editors, the review process, and uses made of the article. However, it is not clear that potential authors will see it in this way. One way to overcome this problem, particularly for small journals that may have to rely more on distribution by separates, is to improve the editorial and reviewing processes. Reduced publishing time would be further incentive to use a particular publisher. Beyond 1985, electronic processes should substantially assist in speeding publication and in producing separates as well. Authors could gain recognition through the current awareness publications and the bibliographic search systems. Publishers that

*Scientists and engineers in academic organizations used electronic equipment about two-thirds as frequently as those from private organizations and four-fifths as frequently as those from government.

distribute separates could maintain records of frequency of requests and citation counts to provide authors with evidence of use of their articles.

THE FUTURE OF PUBLICATION

The growth of science and technology can be expected to yield a corresponding increase in the number of journals published in the United States (Table 10.3). The growth in number of journals is expected to be only about 10 to 20 percent over ten years for the Physical Sciences, Environmental Sciences, and Engineering and a rapid 60 percent for Psychology. Although the growth in number of journals is steady, it is not quite as rapid as in the number of articles because many journals are publishing more articles, a trend that is expected to continue.

The number of subscriptions per journal is also expected to continue to grow through 1985, partly as a result of the increased number of scientists and engineers (Table 10.4). The number of subscriptions per scientist and engineer is anticipated to remain fairly constant, with a moderate increase of 12 percent over a ten-year period. Domestic individual subscriptions are expected to increase about 22 percent from 1975 to 1985, compared to a 17 percent decrease for domestic institutional subscriptions; foreign subscriptions probably will remain at the same level. We expect growth in subscriptions to large journals to be greater than for medium or small journals (21 percent versus –5 percent and 11 percent, respectively), a trend that could portend economic difficulties for small and medium sized journals.

The prices of journals are expected to continue upward to 1985, based partially on the increasing costs of publishing. Prices and typical costs are shown in Table 10.5. Paper costs could increase more rapidly than projected if petroleum costs continue to increase rapidly. Also the projected postal rates could be low since these costs are labor intensive and postal employees are unionized. Future government regulations and subsidization will also play a part.

Electronic processes are currently being used extensively by publishers for editing, redaction, and composition. Direct input from authors to publishers in digital form would reduce the cost of keyboarding even more. Optical character recognition (OCR) might also be used increasingly for direct input. However, the technology and costs of OCR have improved much more slowly than many have anticipated in the past. Composition is being increasingly integrated into general text

Table 10.3   Projected number of journals by field of science, 1975–1985.

Field of Science	1975	1977	1980	1985	Ten-Year Change (%)
Physical Sciences	271	278	284	294	8
Mathematics	98	106	118	138	40
Computer Sciences	62	68	77	92	48
Environmental Sciences	228	234	243	259	14
Engineering	447	471	484	516	15
Life Sciences	1,221	1,318	1,441	1,648	35
Psychology	204	228	263	326	60
Social Sciences	1,340	1,428	1,512	1,678	25
Other Sciences	304	316	331	353	16
All Fields	4,175	4,447	4,753	5,304	27

Table 10.4   Projected average number of subscriptions per journal, total subscriptions, and subscriptions per scientist or engineer, 1975–1985.

	1975	1977	1980	1985	Ten-Year Change (%)
Average number of subscriptions per journal	6,149	6,327	6,384	6,547	6
Total number of subscriptions (million)	25.7	28.1	31.4	36.8	43
Average subscriptions per scientist and engineer	9.8	10.1	10.5	11.0	12

Table 10.5   Projected journal prices and some cost elements, 1975–1985.

Source of Price or Cost	1975	1977	1980	1985	Ten-Year Change (%)
Price					
Individual	$32.08	$32.69	$36.97	$44.28	38
Institutional	34.35	37.72	42.31	49.59	44
Foreign	32.42	42.05	48.98	61.16	87
Cost					
Per-page typesetting (basic text)	$42.40	$47.10	57.00	$66.70	57
Wage index for union printers	7.80	8.30	10.00	12.50	60
Paper cost (per piece)	0.0030	0.0033	0.0035	0.0038	27
Postal rates	0.056	0.060	0.064	0.081	45

processing. Computer-based text-editing systems permit the insertion of operating codes for the electronic composer, a type of integrated system often used by newspaper publishers.[8]

Lerner[9] points out that one obstacle to total computer photocomposition of scientific and technical material has been the problem of setting complex mathematics, or display equations. However, several software packages now provide solutions; ATEX developed one for the American Institute of Physics, the U.S. Government Printing Office has a software package, and Bell Laboratories has its UNIX system, all capable of setting difficult mathematics by using an almost unlimited set of special characters. It is not yet possible to integrate a mathematics page makeup program with illustrations, but in the future one can expect to have full-text computer photocomposition, including illustrations, charts, and figures. Computer photocomposition that can handle complex mathematics is priced competitively with typewriter composition and can be purchased on the open market. Keyboarding of complex mathematics for computer photocomposition is up to 60 percent faster than typewriter composition. Page makeup with galleys from computer photocomposition is easier than page makeup with the oversized galleys produced by typewriter composition. Thus, computer photocomposition has become more efficient and less costly than other forms of composition.

In the future, publishers will have enormous flexibility in the form and mode of distribution of articles made from master images (formated article text). Electronic technology has progressed to the point where articles can be individually printed directly from computer output. Impact printers caused problems with this kind of output, but nonimpact printing has resolved the problem of computer output speed. The ink jet, electrophotographic, and electrostatic systems all hold substantial promise for scientific and technical publishing.

Computer master images can also be transformed into microform or onto video discs. Microform still has many disadvantages: too many distinct formats, reduction ratios, and retrieval coding schemes in use; poor-quality film images and remagnified paper prints; expensive, hard-to-use, and bulky printers; inadequate user environments in libraries; and difficulty in obtaining paper prints.[10] We believe that the greatest promise for microform is with new libraries that want to build up an archival file in certain areas inexpensively. The problem is that the savings in reproduction (runoff) costs with microform are not great enough to overcome the large prerun costs. Thus prices cannot be sufficiently reduced to induce libraries or scientists to purchase this form of distribution.

The new video disc technology holds promise for distribution and storage but it has some weaknesses, since current video images cannot display an adequate full printed page on most inexpensive viewing devices. The strength of video discs is that a very large number of articles can be stored in a small amount of space and then rapidly retrieved. Video discs, written and read by laser beams, can store 108,000 frames on a disc, at 1,000 characters per frame; a form of storage that is very low in cost.[11] The costs of other new forms of information storage are also decreasing. Holographic storage, which is still being developed commercially, offers the advantage of sufficient redundancy of information so that even if a portion of the hologram is destroyed, the remainder can reproduce the whole; holograms on a four-by-six-inch microfiche can store as much as 2×10^8 bits of information in 20,000 holograms. One can even foresee a system that writes and reads information using ion and electron beams. It has been estimated by Richard Feynman[12] that 24 million volumes could be stored in this fashion in the surface area of 7 million pinheads or about three square yards; this would contain all of the information ever recorded in books. If an electron microscope could be built with improved resolving power, it would be possible to store all the books in the world in one pinhead, or in a cube of material only 0.02 inch wide.

Another area in which electronic processes might be employed is in mailing digital tapes, cards, or discs of manuscripts prepared by word processing so that publishers could enter the text directly into a composer. Also the digital text could be transmitted to publishers by telecommunication. Reviewers could be included in a teleconference-like remote access system. The extent of electronic processes used in publishing and their effect on cost is displayed in Table 10.6, which shows that the processes have some cost advantage but not an appreciable one.

Through 1980, the amount of electronic input, either by mail or on-line, is limited by the capabilities of publishers to convert digital word-processing output for electronic text editing and/or composition. Also inability to handle graphics in digital form will hold back this electronic process. Thus it is assumed that there will be only a small proportion of articles in digital form input by mail up to 1980.

Although photocomposition equipment costs have dropped dramatically in the past ten years, some publishers will find it less expensive to continue with fully amortized equipment and others will find it difficult to change because of labor pressures. Our evidence suggests that computer photocomposition should cost about $118 less per article than traditional typesetting, machine costs included. Thus

Table 10.6   Projections of electronic processes used and costs of publication, 1975–1985.

	1975	1977	1980	1985	Ten-Year Change (%)
Articles electronically input by mail (%)			1	15	N.A.
Articles electronically input on-line (%)				1	N.A.
Articles electronically text edited (%)	10	14	20	40	300
Articles with reviewers on-line (%)					N.A.
Electronic composition (%)	20	26	35	60	200
Publishing cost per article ($)	1,490	1,690	1,920	2,210	48
Total publishing cost ($ millions)	483	618	801	1,054	118

the journals that used electronic composition in 1975 saved about $7.7 million. Preparation of articles by electronic text editing will save an additional $9 per article.

The few data available suggest that these innovations are much more prevalent with large circulation journals than with small journals. We estimate from our publisher survey* that about 10 percent of the small journals, 30 percent of the medium journals, and 40 percent of the large journals are electronically photocomposed. This means that because small journals have less access to the capital needed for electronic photocomposition, they also have less access to the savings that can be achieved in this manner. In 1980, it is estimated that the typesetting costs of small journals will account for 37 percent of their total prerun and runoff costs. On the other hand, the typesetting costs of large journals will account for 15 percent of their total cost. The differences in number of subscribers and in prerun costs result in much higher prices for small journals.

Since fewer small than large journals will have electronic processes available, publishers of small journals will presumably seek other ways to reduce prerun costs. More of them will print directly from typed input, and others may attempt to reduce costs by having less reviewing performed. It is also expected that page charges will become more prevalent and that their amount will increase to cover more of the prerun costs. Some publishers will try other innovations—publishing in

*The response rate on this survey was only 28 percent; thus some caution must be taken in interpreting the results.

microform, publishing smaller articles, or publishing synoptic journal articles.

In some instances, publishers of small journals may increase their use of electronic processes by acquiring equipment, sharing facilities (such as with an editorial processing center), or subcontracting these processes out to appropriate facilities. A federal ruling is anticipated that will preclude journals, which levy page charges paid directly or indirectly from federal grant or contract funds, from charging royalty payments under copyright. This may account for about 40 percent of the articles published. More journals will begin to adopt page charge policies, and the amount of the fees will be increased in some cases to cover all prerun handling costs, actions that will help in keeping subscription prices down.

We project that by 1985 nearly 60 percent of the journals will be composed by electronic processes, and about 40 percent of the manuscripts will be edited in this way. Many of the barriers to change will be overcome by equipment obsolescence, labor retirement or change, widespread use by other publishers, and reduced costs. The saving of the expected use of electronic processes is about $61 million, with about $49 million attributable to electronic composition, $9 million to telecommunication input, and $2.5 million to text editing. The saving for complete electronic publishing is $370 per article.

Some publishers may publish in microform for distribution by 1985. However, the principal use of this form will probably be for back issues purchased by new libraries or libraries that begin new holdings. Libraries that change locations may even find that purchasing this storage medium is less expensive than moving an entire collection. Microform equipment will be available to most academic and public libraries since they are already frequently used for other purposes.

We expect that the reduced prerun costs plus income from other sources will help keep prices of journals in line with past trends, with some minor reductions possible from forecast prices. Journal publishers are furthermore expected to continue vigorous pursuit of other sources of income. The amount of advertising is likely to rise in large circulation journals. Page charges will increase both in terms of the proportion of journals using them and in the amounts requested. Sales of reprints to authors will probably increase. Sales of separates by publishers are expected to increase modestly, up to 1 million copies by 1985.

Without improvements from electronic processes, beyond 1985 many journals would be under even stronger pressure to find other means of reducing costs, which could hurt quality, or they might seek other sources of income, such as royalty payments, advertising, sale of reprints, and page charges. Advertising is an unlikely source of sufficient income to small journals; not many advertisers are interested in targeting advertising to the highly specific audiences found with small journals. Sale of reprints can help, but our data suggest that this would compete with subscription sales by providing another, albeit inefficient, source of articles. Page charges may also be a good source of income. However, if large journals charge less because of lower prerun costs, the number of articles submitted to small journals may dwindle due to competition and costs. Also commercial publishers cannot be paid page charges from federal grant or contract funds. Thus none of these additional sources of income seems promising.

Possibilities of reducing costs include publishing in microform or on video discs, reducing the size of journal articles, reducing editorial or review quality, and using typed manuscripts as input to reproduction. Nearly all of these possibilities are now being tried by some journals or are discussed as innovations that might be attempted in the near future.

We feel that most of the text editing, redaction, and composition beyond 1985 will be done by electronic processes. We also believe that most of the input from authors will be in digital form, since manuscripts are progressively more available in that form. Telecommunication will possibly be used for transmission for most of the manuscripts because these costs for digital data will be very low.

Distribution from publishers is expected to be in traditional paper form, as well as digital form and microform. Many journals with large circulation could provide all three forms, with the paper form dominating for economic and efficiency reasons. Most small journals will also distribute copies in all ways, but digital forms could become much more prevalent. Publishers of these journals will rely almost entirely on income from sales of separates, page charges (or else copyright royalties), subscriptions to current awareness announcements, and sale of abstracts, indexes, and citations to bibliographic data-base distributors and other secondary services. Most of these articles will be transmitted in digital form. It is anticipated that nearly all of these articles will be stored somewhere in digital form for on-demand separates distribution. Individual subscriptions are expected to drop to nearly one-half the number that would be expected if we extrapolated by trend. Distribution by preprints and reprints is expected to dwindle to about 20 percent of the current number and much of this will be through teleconferencing networks. We feel that nearly two-thirds of the uses through libraries will be in the form of separates requested from a national system or individual publishers, with about half of the transmissions made by telecommunication and half from digital storage but mailed in paper form or microform.

THE FUTURE OF LIBRARIES AND SECONDARY ORGANIZATIONS

The level of growth of the number of libraries is expected to continue through 1985; however, the patterns of journal purchases will not necessarily follow that growth. Also new technology may make some large differences in the role libraries will play in the future journal system. The growth of libraries and their relative journal purchases is given in Table 10.7. The average number of library subscriptions is expected to drop, partly because estimated journal expenditures are rising faster than the estimated total scientific and technical material expenditures. This result is in line with those made in the Indiana University study.[13] Although there is no direct evidence, libraries may be substituting interlibrary loans for purchases of journal subscriptions. Certainly the number of interlibrary loans is high (5.3 million in 1977) and is increasing.[14,15]

One of the manifestations of increased interlibrary loans is a plan for a new system for handling future interlibrary loan needs. The National Commission on Libraries and Information Science recently endorsed a report that strongly recommended that a formal national periodicals system be implemented in the United States by 1981.[16] The system is based on a loosely defined hierarchical network of libraries. The focal point of the system would be a National Periodicals Center, operated by the Library of Congress or some other organization. This center would concentrate on titles currently published and back files thereof. Requests for defunct titles that cannot be located would be obtained through agreements with such organizations as the British Library Lending Division. Only about 20 percent of the total loans are expected to come through the center. The remainder would be handled by individual libraries or through state or regional systems. There are also other periodical systems now in operation or being proposed. For example, photocopies, reprints, and microform copies of journal articles can be requested from such organizations as the Institute for Scientific Information, Congressional Information Service, and University Microforms International.

If the Copyright Clearance Center (CCC) is successful, it too could serve as a central facility for distributing copies of articles. The CCC has gained some acceptance from publishers but not from many libraries. If this barrier can be overcome, its handling of photocopying transactions and accounting for royalty payments could make the CCC a candidate for the central facility. Also organizations such as the Center for Research Libraries, Institute for Scientific Information, and University Microforms International are all handling requests for copies of journal

Table 10.7 Numbers of libraries, subscriptions, and expenditures, 1975–1985.

	1975	1977	1980	1985	Ten-Year Change (%)
Number of libraries (000)	20.3	23.0	24.3	29.7	46
Total number of library subscriptions (millions)	8.0	8.0	8.1	8.6	8
Average number of subscriptions	394	348	333	290	−26
Estimated total S&T material expenditures ($ millions)	215	239	279	349	62
Estimated total journal expenditures ($ millions)	380	433	551	800	110

Table 10.8 Library cost elements, 1975–1985.

	1975	1977	1980	1985	Ten-Year Change (%)
Average number of library subscriptions per journal	1,922	1,789	1,736	1,612	−16
Number of interlibrary loans (millions)	3.3	4.0	5.8	7.8	136
National Periodicals System (millions)					
Total loans			0.5	3.0	N.A.
Paper storage and distribution			0.5	1.0	N.A.
Electronic storage, paper distribution				1.9	N.A.
Electronic storage and distribution				0.1	N.A.

articles and could expand their service as a central facility. The interest shown by the public and private sectors in distributing copies of articles suggests that the time may be near for some form of a national periodicals system.

Some of the library cost elements for a national system are given in Table 10.8. The average domestic institutional subscription rate is forecast to move down from about 1,900 subscriptions in 1975 to about 1,600 in 1980. Many of the canceled subscriptions are expected to be of small journals, which are infrequently read in many libraries. Libraries that cancel subscriptions will depend to an increasing degree on interlibrary loans for obtaining copies of articles from journals. The new copyright law may contribute to the number of canceled subscriptions because libraries with fixed budgets may reallocate some of their subscription budgets to royalty payments and fees for use of the Copyright Clearance Center.[17] The journals that lose library subscriptions will often not receive compensation from royalty payments due to the CONTU guidelines,[18] which state, in effect, that the borrowing library is responsible for royalty payments for more than five photocopies made of publications fewer than six years old (excluding photocopies made under fair-use provisions). The cost per interlibrary loan to both borrowing and lending libraries is estimated to average $23 in 1980. Some lending libraries will charge for the service, dampening the number of loans somewhat. A modest number of libraries will probably purchase journal copies in microform in lieu of paper copy. However, most of these would be new libraries acquiring collections of older materials. For the most part, individuals and libraries will continue to acquire and store copies of journals in the traditional manner. Some costs will be saved in libraries by not binding old issues of journals and by better methods of compact storage. The use of video discs may also become prevalent in libraries, thereby expanding their coverage and reducing storage requirements.

It is anticipated that about 500,000 requested copies will be handled by the National Periodicals Center (NPC) in the early 1980s from among the 5.8 million interlibrary loans. In order for the NPC to gain acceptance by libraries, it must be able to handle requests for at least some older journals because so many interlibrary loan requests involve these. The NPC must also provide royalty payments for the publishers that desire these payments. Not all publishers expect such payments, however.[19] Small journals, in particular, will benefit from having this facility available for distribution of separate article copies. Some of them may even prohibit photocopying of articles by one library for another as a condition of the library subscription price.

The royalty payment requested by small journals can be anticipated to approximate the average cost per use where all uses come through distribution of separates. This means that the royalty payments could be as high as $5 per article for some small journals. The cost of the NPC central facility is expected to be about the same as interlibrary loans in the early 1980s because start-up and acquisition costs will approximately balance savings from more efficient operations. The requesting libraries are expected to pay a fee for the NPC service, which includes sufficient royalty payment to the journal publishers. The average 1980 cost of interlibrary loans, exclusive of royalty, is $13 for the borrowing library and $10 for the lending library. The comparable operating costs are about $3 for the requesting library plus $7 for the NPC, which is passed on to the requesting library. The cost to the requesting library is less when the NPC is used because there is no need to look up possible sources of the article, and the necessity for duplication of requests to more than one lending library is eliminated.

Interlibrary loans not processed by the NPC are projected to be about 5.8 million article photocopies in the early 1980s. The NPC is not likely to have much effect on the number of intrasystem loans, estimated at 12.5 million articles in the early 1980s. Unless it becomes less expensive to process requests through the national system, local central libraries will continue to perform this operation for their branch libraries.

One of the areas in which new technology has been directly applicable to scientific and technical communication is in library operations and services. The most prominent of these are computer-based (on- and off-line) bibliographic searches, automated circulation, cataloging of books, and interlibrary loan processing. One technology outside of scientific and technical communication that might be important is mass storage memories. If the NPC comes into being, one component might be digital storage of input directly from publishers. Thus far, the most likely is magnetic tape segmented for mechanical handling. However, some output problems associated with mass storage memories are in queuing delays, and the output costs are prohibitively high. Other barriers to mass electronic storage are the needs for costly, labor-intensive indexing and control and more effective access software.

Abstracting and indexing (A&I) services play an important role in providing the intellectual analysis and organization of the scientific literature and thus allowing effective access to articles by scientists and engineers engaged in research activities. While concentrating primarily on the journal literature, A&I services treat all literature sources to some extent and provide scientists with current awareness of materials relevant to their fields of interest as well as guides for retrospective search of the world's literature. The traditional product of A&I services is a periodic publication listing items, often with abstracts, and frequently accompanied by extensive indexes. In the development of these products, a number of organizations have created machine-readable data bases, which can be used to provide services directly. The growth of such services has been substantial in recent years, suggesting correspondingly less emphasis on the printed A&I product in the future.

About 330 U.S. A&I services covered areas of science and technology in 1975. Most of the major organizations are members of the National Federation of Abstracting and Indexing Services (NFAIS), which estimated that U.S. A&I services processed about 2.9 million scientific and technical items in 1975, up from 1 million in 1960.[20] This number includes a considerable proportion of items published outside the United States and some overlapping coverage among the different services.

Another information product is bibliographic data bases. Initially created as by-products of the A&I publication process, these are increasingly becoming an information product in their own right. Covering in general the same literature as do the traditional products, they allow in-depth searching of large volumes of information stored by computer. As the development of bibliographic data bases has increased, related service organizations have been established to facilitate their use. These include processors and suppliers, who make the data available, and brokers, who perform computer searches for users.

The accumulation of stored materials available to scientists and engineers in libraries is expected to continue to accelerate. Furthermore more research will be interdisciplinary so individual subscriptions will become less useful for obtaining wanted articles. Some of the pressures of identifying articles will be eased by on-line retrieval systems. The forecast of the number of on-line bibliographic searches is 2.5 million in 1980, up from 1 million in 1975. However, some libraries are expected to cancel subscriptions to abstract journals by 1980 and to substitute on-line searches for these subscriptions. This will exert substantial economic pressure on A&I services since the income from on-line searching appears to be not nearly as great as from the subscriptions they displace.

It is anticipated that not many libraries will subscribe to microform journals because the price will not be much less than for paper copies and the advantages do not outweigh the disadvantages. Identification of appropriate articles may become even more difficult by 1985 due to the size of accumulated stores of information, immediate availability of fewer copies, and increased interdisciplinary research. On-line searches are expected to increase to 3.7 million in 1985.* However, any growth may be hampered by the fact that their prices will be increased to help recover the large input costs, which can no longer be completely recovered from the many fewer paper copy subscriptions to abstracts and indexes.

Major changes are anticipated by 1985, particularly in the area of distribution of separates through author reprints, interlibrary loans, and the NPC. We anticipate that 3 million of the article copies estimated to come from authors will now come from publishers and the center. Some publishers are expected to begin to distribute separates on request in order to make up for income lost from reduced library subscriptions.

---

*Estimates available at the time of publication of this book indicate that the number of on-line searches had already surpassed 3.7 million by 1979. This demonstrates one of the dangers of forecasting in a rapidly changing area.

This number is expected to be about 1 million copies in 1985. Since these articles will be in digital form, the expenses of storing and retrieving requested copies will be less, although the cost of reproduction must be added. Some of the publishers can be expected to have the capability to retrieve directly and to print composed copies by electronic processes.

About 3 million journal copies will be provided by the NPC in 1985. The anticipated increase in the use of the NPC is attributable to a rise in the number of lending libraries that charge for interlibrary loans and to the lower costs of the NPC. By this time the center will have holdings of virtually all U.S. technical periodicals five years old and some further back. The remainder of older requests will be referred to other facilities. Some publishers may by then transmit their articles (estimated at 80,000) to the NPC in digital form by mail. The digital input will be handled by computer, and hard copies may be printed from computer output and mailed to requesters. The digital storage of articles may be in a mass storage devices or merely by individually stored tapes or floppy discs.

Since some requesters will have intelligent terminals or minicomputers and communication modems for transmission at 2400 baud or above, they can request and receive full text of articles by telecommunication. The number of such requests, however, is expected to be less than 100,000 in 1985 unless some major breakthroughs occur in computer output costs, transmission costs, and technology for digital input and output of special graphics. We expect that most electronic input and transmission of articles will be from journals that have few graphics. Such journals are found to be more prevalent in Psychology and some of the Life Sciences.

The cost of an interlibrary loan is estimated to be about $29 in 1985, including costs to both lending and borrowing institutions. The cost to the lending library is estimated to be $12. On the other hand, the cost of using the NPC is $8 for requests fulfilled from electronic storage and transmitted in paper copy (1.9 million) and $6–10 for requests fulfilled from electronic storage and transmitted electronically (100,000).

In 1985, many of the requests to the NPC will probably be for articles from small journals, whose publishers could begin to realize that such distribution provides an essential portion of their income. They will charge royalties accordingly, which might be well over $5 per article depending on the amount of use. The increased use of the NPC may also begin to displace some use of preprints and reprints obtained from authors by 1985. Part of the reason for this is that people will become accustomed to NPC services, and it should be more efficient and substantially faster, although more expensive, to obtain copies of articles through the system than from authors.

The extensive use of separates, coupled with reduced transmission costs, would make the use of on-line searches of bibliographic data bases essential. Also publishers will undoubtedly begin to publish current awareness announcements for their articles. They will prepare most of the bibliographic entries and some of the abstracts for the A&I data bases. These entires can be transmitted to secondary organizations in digital form for ease of handling and speed of transmission, thus reducing publishing delays.

Beyond 1985, the number of library subscriptions to journals is expected to drop dramatically. Author reprints are expected to survive at about one-third of their current level of production. A very large proportion of the library and author reprint uses are expected to have been displaced by publishers' distribution of separates and the NPC. By this time the center would have ten to twenty years' worth of accumulated digital storage for a large proportion of the journals, stored individually or in a mass storage system. Either way, the speedy response to libraries and end users will make the system a highly desirable one. It is expected that over one-fourth of the transmissions from publishers and the National Periodicals System will be by telecommunication since costs will be low and the technological breakthroughs necessary to permit digital transmission of graphics as well as text can be anticipated by then.

THE FUTURE OF JOURNAL USE

The extent of journal reading is expected to continue rising at about the same rate as the number of scientists and engineers. We feel that the average number of readings per scientist or engineer will not change much, although the patterns of reading may. There appears to be a trend toward more and more articles being read not in bound issues but as individual copies: author preprints or reprints, interlibrary loan photocopies, publisher reprints, or copies from services such as those offered by the Center for Research Libraries, the Institute for Scientific Information, and the British Library Lending Division. Part of this trend is due to the availability of the separates through these services. Other reasons include the proliferation of interlibrary loan and referral services such as Minnesota Interlibrary Telecommunication Exchange (MINITEX) in Minnesota and New York State Interlibrary Loan Service (NYSILL) in New York. The on-line retrieval systems must surely be contributing to this trend as well, since readers can now extend their means of identifying needed information far back over the years as well as to a breadth not thought possible twenty years ago. Finally the multidisciplinary nature of current science and technology may lead to the requirements for new sources of information beyond one's own journals that are obtained through professional societies.

Technology will probably have a less immediate effect on journal readers than on the other participants because their function lends itself less to such technology. However, in the long run it is likely that users will begin to receive copies of article text by electronic communication, principally by digital output and video displays. Scientists and engineers can use these devices for full-text retrieval and teleconferences, as well as for on-line bibliographic searches performed prior to full-text retrieval. The utility and success of such electronic processes depends to a large degree on telecommunications access and cost. More economical telecommunication will come through new technology such as fiber optics and sending and receiving terminals, which will permit storage of received messages while awaiting printout or display. Then the actual communication channels will be used for a much briefer time, which will reduce the communication costs drastically.

The number of readings and the sources of articles are given in Table 10.9. The total number of uses of journal articles is expected to increase from 231 million in 1975 to 263 million in 1980. Most of the uses are expected to continue to come from individual subscriptions (192 million), with the rest from institutional subscriptions (39 million) and author preprints and reprints (32 million). The number of copies of articles obtained from separates are projected to increase proportionately because reprints, preprints, interlibrary loans, and intrasystem loans are all expected to go up proportionately. Acquisition of separate copies will continue to be inefficient from the standpoint of identification, locating sources (authors in particular),* speed of transmission, and costs. The total number of article uses is anticipated to increase to 294 million in 1985 due to increases in the numbers of scientists and engineers. Most of the uses will still be from individual subscriptions (214 million), with institutional subscriptions and authors' preprints and reprints accounting for 44 million and 36 million uses, respectively. The total number of interlibrary loans is projected to be 7.8 million. About 1 million of the separates are assumed to come from reprints distributed by publishers and 3 million from some form of an NPC.

Beyond 1985, on-line bibliographic searching and teleconferencing by scientists and engineers will provide a climate in which on-line

*One or more address change by an author since publication is common.

Table 10.9 Projected number of readings and the sources of articles, 1975–1985 (millions).

Sources of Articles	1975	1977	1980	1985	Ten-Year Change (%)
Total number of readings	231	244	263	294	27
Individual subscriptions	167	178	192	214	78
Institutional subscriptions	34	36	39	44	26
Interlibrary loans	3.3	4.0	5.3	7.8	136
NPC			0.5	3.0	N.A.
Author preprints/reprints	30	30	32	36	20
Publisher reprints	0.6	0.6	0.7	0.9	50
Cost per use	$15.86	$18.59	$22.40	$28.50	79
Total use cost	$3,670	$4,535	$5,880	$8,370	128

retrieval of the full text of articles will be commonplace. Lower telecommunication costs and better direct access to digital storage are expected to foster the use of this mechanism.

The average cost per use is estimated to be $28.47 in 1985, an increase of 80 percent since 1975. The total use cost is estimated to increase from $3.7 billion in 1975 to $8.4 billion in 1985, a 128 percent increase. It would require a well-articulated technological-economic model plus some major assumptions to assess the effect of technology on cost of the use function; no attempt is made to do so here.

THE FUTURE OF JOURNAL TRANSMISSION

Transmission is an important function. Currently journal transmission is accomplished by mailing or personally transferring manuscripts or articles. This function is characterized by a very large number of individual transmissions, which do not cost a great deal in relation to other costs but are slow. In fact, rapid transmission may be the most important electronic process in the future. At this time very little, if any, transmission of the full text of articles is performed electronically. Exceptions are the one existing electronic journal in computer technology and occasional articles sent by telefacsimile. In addition, some messages about articles, such as on-line bibliographic data, library cataloging, and requests for interlibrary loans, are sent by telecommunication.

We estimate that in 1975 about 2 million draft manuscript copies were transmitted among coauthors or from authors to colleagues to obtain critiques. Another 1 million manuscript transmissions occurred between authors and publishers, with exchanges involving revisions, page proofs, and rejected manuscripts. An estimated 1.5 million transmissions were between editors and reviewers. Perhaps 2.5 million letters were transmitted among editors, authors, and reviewers. A total of about 80 million issues of journals were sent to individuals and another 52 million issues to libraries. These two groups of transmissions involved mailing a total of over 2.5 billion individual copies of articles. There were about 10 million interlibrary loans that involved two-way transmission for request and fulfillment. About 46 million reprints were sent from publishers to authors and then to users. Finally there were about 200 million letters of requests for subscriptions, interlibrary loans, reprints, and so on. It is estimated that these transmissions cost about $700,000 or $0.10 per use.

Most telecommunication currently is by voice grade telephone lines. Moreover, value-added networks provide a potentially low-cost telecommunication capability for all participants in scientific and technical communication. The potential is enhanced by the availability of minicomputers or intelligent terminals, which permit rapid transmittal and buffering of messages. New technologies in long-line communication such as fiber optics, digital transmission and switching equipment, and communication satellites could yield substantial decreases in average cost provided that the capital cost is shared by enough users (not restricted to scientific and technical information communication).

Prestel (previously called Viewdata) is a new communications system developed by the British Post Office; it links television sets with a centralized computer data base by means of normal telephone lines, thus allowing individuals, commercial concerns, and public institutions such as libraries immediate access to a large and constantly expanding source of information. The system allows users to select the information they require and to respond to information presented on the television screen. Thus this system could permit searching and retrieval of journal articles to be performed in the home as well as offices and libraries. The progress of this system in England and in the United States is well worth following, along with systems such as Teletext, the U.S. Department of Agriculture's "Green Thumb" system, and the synthesized-data systems currently being developed and tested by the National Library of Medicine/Lister Hill National Center for Biomedical Communication.

In 1980 most article manuscripts will still be mailed from authors to publishers and all communications among authors, publishers, editors, and reviewers will be by mail, whether in the form of manuscripts, page proofs, or letters. However, since many more manuscripts are expected to be prepared by electronic processes, we anticipate that some of them will be mailed in digital form to reduce keyboarding time and expense.

Only a small number of publishers are expected to have recognized and to have acted on the real advantage of having manuscripts that can be entered directly into photocomposition without rekeyboarding. Some journal publishers will accept OCR-readable manuscripts while a lesser number will accept magnetic tapes or floppy discs. Even in 1985, a substantial proportion of these communications are assumed to be by mail. On the other hand, many of the publishers are expected to accept OCR-readable or digital input in order to avoid keyboarding costs. It is expected that as much as 15 percent of the manuscripts will be sent by mail in digital form. The principal barrier to this form of input is that some equipment and programs used by publishers will not be compatible with the digital forms prepared by author word-processing or text-editing systems. We expect no adequate standards or conventions to be available by 1985 to make these electronic processes completely compatible. Since some authors and some publishers will have intelligent terminals or minicomputers available, some telecommunication of text will occur (perhaps 1 percent). With appropriate equipment, the transmission costs could be as low as $1 to $2 per article. The ease of transmission and decreased publishing time (about one to two weeks saved), would make this option desirable to some authors and some publishers.

Beyond 1985 it is expected that some members of all participant groups will use electronic processes. Thus it may be possible to process and to communicate many articles entirely electronically. As many as 50 percent of the manuscripts will be sent to publishers in digital form, with 10 percent by telecommunication, and 5 percent will be telecommunicated to reviewers. Many separate copies of articles are also likely to be sent by telecommunication. It is anticipated that many separate copies will be sent by telecommunication from publishers or an NPC to libraries or end users because most of the articles will be in digital form, many of the senders and receivers will have intelligent terminals or minicomputers for accommodating high speed transmission, and the telecommunication costs will probably have decreased. This means that many articles will be communicated entirely by electronic processes; little paper will be involved except to the extent that users may prefer that form of output. Some of these articles may be from small journals that distribute no paper copies at all. However, some will also be from journals that distribute articles in multiple forms (digital, paper, and microform) depending on the economics of each

Table 10.10 Projected journal costs and cost per use, 1975–1985.

	1975	1977	1980	1985	Ten-Year Increase (%)
Total cost (millions)					
All uses	$3,670	$4,535	$5,880	$8,370	128
Individual subscriptions and publisher reprints	2,312	2,875	3,737	5,333	131
Institutional subscriptions	915	1,046	1,347	1,914	109
Author preprints/reprints	478	572	744	1,074	125
Cost per use					
All uses	15.90	18.59	22.40	28.50	79
Individual subscriptions and publisher reprints	13.80	16.10	19.50	24.80	80
Institutional subscriptions	26.80	29.00	34.70	44.00	64
Author preprints/reprints	15.90	19.40	23.50	30.20	90

and the equipment capabilities and desires of the intermediary participants and end users.

THE FUTURE COST OF THE JOURNAL SYSTEM

In each of the previous sections of this chapter constituent costs were estimated. Table 10.10 summarizes these costs. The cost of the journal system is expected to rise rapidly, but the cost per use will not increase as rapidly due partly to advanced electronic processes used throughout the journal system.

A COMPREHENSIVE ELECTRONIC JOURNAL SYSTEM

Recent technological advances, which were developed largely independently of scientific and technical communication, provide all of the components of a comprehensive electronic journal system. Such a system would provide enormous flexibility, particularly because individual articles can be distributed in the most economically advantageous manner. Much-read articles may still be distributed in paper form, and infrequently read articles can be requested and quickly received by telecommunication when they are needed. The trade-off is that resources formerly wasted in printing, mailing, and storage would be applied to better identification and retrieval of information, thus reducing cost, improving quality, and increasing efficiency. Other benefits of electronic processing can also be derived. Furthermore, we believe, better systems integration will yield more emphasis on the quality of articles' content, less republication of articles for updating or for different audiences, and better access and retrieval to information needed in multidisciplinary research.

In the electronic journal system, articles will be prepared by authors using sophisticated text-editing systems. Article preparation may include joint writing of text through teleconferencing systems in which immediate peer review is possible, comments are made, and specific research questions can be answered. Many of the citations used will come from those found in on-line bibliographic searches. When citations are identified they can be immediately retrieved by telecommunication in full text on cathode ray tube (CRT) or in paper form. The digital form of the unreviewed manuscript will be directly transmitted electronically to a publisher. The publisher will electronically transmit the manuscript to a subject editor, who will read the text by CRT or printout and make electronic notes concerning editorial and content quality. The subject editor may choose appropriate reviewers using a computer program that matches the profile of potential reviewers with the topics covered in the article. Other computer-stored information will be used to help screen reviewers, such as by affiliation and relationship to the authors, status of the most recent review, frequency of reviews, timeliness of response of previous reviews, and quality of reviews. The reviewers will respond to editors, and editors in turn to authors, by telecommunication, comparable to current teleconferencing processes. Publishers and editors can also use electronic processes for business purposes, address listings, and other such activities.

An accepted article will be subject to redaction on a text-editing terminal, and the computer-based text will be provided in several forms, including full text and bibliographic description. The bibliographic description will be transmitted electronically and used directly by search services or will be entered into abstracting and indexing services for further analysis and processing. The full text will be sent electronically to some individual scientists designated by the author or by request to scientists based on its topic, author, or other bibliographic identifier.* The scientists may receive the text on personal terminals or on library terminals. Articles will also be sent directly by telecommunication to the

*A combination of the author's automatic reprint distribution and users' SDI service.

NPC, which will have a central archive and several decentralized computer stores. The NPC will await telecommunicated requests for copies of articles and will respond with full text telecommunicated to the requesters.

The electronic processes also provide a great deal of flexibility of output, which can enhance reading and assimilation of the information. For example, end users could request alternative formats of the text to suit their particular needs, perhaps for rapid scanning or for in-depth reading. Rapid scanning can be facilitated by highlighting certain elements of text, narrowing column widths, widening spaces between lines, and other techniques. Human factors modifications can also help in-depth reading by design and page layout and by highlighting certain terms. Electronic processes can aid in combining mathematical formulas, data presentations, and text in a way that meets alternative needs.

This comprehensive electronic journal system is highly desirable and currently achievable. It is believed that within the next twenty years, a majority of articles will be handled by at least some electronic processes throughout but not all articles will be incorporated into a comprehensive electronic journal system. Some articles will be processed electronically in different ways depending on the electronic capabilities of the senders and receivers involved.

CONSTRAINTS ON AN ELECTRONIC JOURNAL SYSTEM

One of the principal constraints to any alternative communication system is the lack of incentive for the participants to change. For example, authors publish partially for prestige and recognition, which results in professional advancement. Certainly the "publish or perish" environment that exists in some fields of science and in some organizations creates incentive to publish, and therefore any alternative communication system must meet this perceived need. Many publishers lack a financial incentive for drastically deviating from the current journal publishing practices. While many book and small journal publishers appear to have financial problems, most large publishers are doing very well financially. They earn a comfortable margin on income and require much less capital to publish journals than books since the income from subscriptions is received before most costs are incurred. Substitution of royalty payments for subscription income will lessen this advantage since photocopying takes place over a long time so the current value of royalty income is less than the value of an equal amount of subscription income. Thus any new publishing systems must incorporate some financial incentives or publishers are unlikely to be interested. Another problem is that some income is derived by publishers through sale of advertising. Distribution of separates appears to preclude advertising. However, Whitby[21] has suggested that advertising can be focused on specific scientists by advertising interest profiles and can then itself be distributed as separates. This may be much better than current advertising practices.

Scientists as users present some barriers to new systems.[22] It has been claimed for many years that scientists would quickly adapt to the direct use of computer terminals and would search bibliographic data bases on-line. However, most have no easy access to terminals or are reluctant to use them so still rely on an intermediary to perform their searches.[23] Regardless of the reasons, if an alternative journal publishing and distribution system requires direct on-line communication, some incentives to use it must be provided to scientists, and their behavior must be altered. We believe that in the future, new scientists who have been trained on terminals in high schools and universities will find it unacceptable not to have these facilities available for analysis, text processing, search, retrieval, and other forms of communication.

Libraries have been in the forefront of adopting on-line searching as an information tool, but they often have little incentive to modify their mode of operating unless their patrons and funders desire such change. Many libraries currently are automating for cataloging and internal recordkeeping, but they still require motivation to change their procedures in dealing with scientists. Some outside incentives will probably be necessary.

Other constraints are technological. Standards must be set for word-processing and text-editing output so that publishers can receive it and easily convert it to the appropriate format. A major problem exists in treating nontextual input—tables, mathematical formulas, and graphics. Technologically, graphics can be electronically handled now, but the economics are not practical for the high volume of graphics found in Physical Sciences, Engineering, and Life Sciences articles. Another requirement is that the cost of telecommunication must continue to fall. Unfortunately, there is some indication that Federal Communications Commission rulings could damage this prospect by restricting free competition and by permitting local line charges to increase.[24] Sending and receiving equipment must be sufficiently sophisticated to permit rapid, and therefore low-cost, communication. Mass storage devices now available could economically store nearly all current literature, but the cost of input and output will remain unacceptable until some breakthroughs in this area are made.

The opportunities for electronic journal processing, transmission, and access are substantial—the opportunity for an author to preselect a specific target audience, for example—but the number of questions to be answered are great. For example, are electronic communication processes better suited for popular entertainment and communication, or for economic transactions such as those envisaged for electronic funds transfer systems? Or is scientific communication such a complex economic and sociological phenomenon that progress will be hindered by tradition, as in any other social system involving complex interactions among numerous participants? We are optimistic. Already our children's electronic toys are more sophisticated than the transistor radios of only twenty years ago, and our children are the scientists of tomorrow.

REFERENCES

1. V. E. Palmour, M. C. Bellassai, and L. M. Gray, *Access to Periodical Resources: A National Plan* (Washington, D.C.: Association of Research Libraries, 1974).
2. Task Force on a National Periodicals System, *Effective Access to the Periodical Literature: A National Program* (Washington, D.C.: National Commission on Libraries and Information Science, 1977).
3. D. W. King and N. K. Roderer, *Systems Analysis of Scientific and Technical Communication in the United States: The Electronic Alternative to Communication Through Paper-Based Journals* (Rockville, Maryland: King Research, 1978).
4. R. D. Salzman, "The Computer Terminal Industry: A Forecast," *Datamation* 21(11):46–50 (1975).
5. D. Ness, "Office Automation Today and Tomorrow," *EDUCOM* 12(3):2–12 (1977).
6. R. B. White, "A Prototype for the Automated Office," *Datamation* 23(4):83–90 (1977).
7. L. A. Green and S. T. Hill, *Editorial Processing Centers: A Study to Determine Economic and Technical Feasibility, Part IV: Survey of Authors, Reviewers, and Subscribers to Journals in the Life Sciences* (NSF - C769) (Rockville, Maryland: Westat, Inc. and Aspen Systems Corporation, 1974).
8. Council on Library Resources, *A National Periodicals Center: Technical Development Plan* (Washington, D.C.: Council on Library Resources, 1978).
9. R. Lerner, "Electronic Publishing" (Paper presented at the 39th Annual Congress of the FID, Edinburgh, September 1978).
10. R. L. Wigington, "Introducing New Technology" (Paper presented to the Engineering Foundation Conference, Rindge, New Hampshire, August 14–19, 1977).

11. Lerner, "Electronic Publishing."
12. Ibid.
13. B. M. Fry and H. S. White, *Publishers and Libraries: A Study of Scholarly and Research Journals* (Lexington, Massachusetts: Lexington Books, 1976).
14. King Research, *Library Photocopying in the United States: With Implications for the Development of a Copyright Royalty Payment Mechanism* (Washington, D.C.: Government Printing Office, 1977).
15. Task Force on a National Periodicals System, *Effective Access.*
16. Council on Library Resources, *National Periodicals Center.*
17. King Research, *Library Photocopying.*
18. U.S. Congress, *Congressional Record House,* "Joint Explanatory Statement of the Committee of Conference," pp. H11 728-H11 729 (September 29, 1976).
19. B. M. Fry, H. S. White, and E. L. Johnson, "Scholarly and Research Journals: Survey of Publisher Practices and Present Attitudes on Authorized Journal Article Copying and Licensing" (Bloomington, Indiana: Graduate Library School, Indiana University, 1977).
20. National Federation of Abstracting and Indexing Services, *Member Source Statistics* (Philadelphia, Pennsylvania: National Federation of Abstracting and Indexing Services, 1974–1976).
21. O. W. Whitby, "Computer Architecture for External Editorial Processing," *Journal of Research Communication Studies* 2:9–23 (1979).
22. D. W. King and V. E. Palmour, "User Behavior," in *Changing Patterns in Information Retrieval,* ed. C. Fenichel (Philadelphia, Pennsylvania: American Society for Information Science, 1974).
23. R. V. Katter and D. B. McCarn, "AIM-TWX: An Experimental On-Line Bibliographic Retrieval System," in *Interactive Bibliographic Search: The User/Computer Interface,* ed. D. E. Walker (Montvale, New Jersey: AFIPS Press, 1971).
24. M. Gerla, "New Line Tariffs and Their Impact on Network Design," *AFIPS Conference Proceedings* 43:577–582 (1974).

Index

Abstracting and indexing
services
    costs, 219–220
    coverage, 148–150
    definition, 146–147
    directories, 147–148
    future, 300–303
    number, 148–150
    usage, 34–35
Acceptance rates, 67–68
Advertising income, 237
Age of articles, 172
Age of authors, 57
Authors
    activities, 66, 192
    age, 57
    costs of authorship, 65–67,
        75–78, 218–219
    female, 57–58
    foreign, 200–206
    funding, 58–59
    selection of journals, 62–65
    time spent writing, 21, 66–67,
        76
    types of, 55

Books, 25–26
Browsing, 165
Bundling, 230

Citations, 34–35, 167, 170–173,
    200
Copyright, 38–41, 143, 276–284,
    297–300
Cost measures, 208–209

Cost model, 109–127
Costs
    abstracting and indexing
        services, 214–220
    authorship, 65–67, 75–78,
        218–219
    communication path, 221–223
    data bases, 155–157
    editing, 110–111, 117–118
    field of science, 223
    fixed, 211–212, 232
    per journal, 97
    libraries, 25–37, 114–146, 219
    marginal, 238–239
    measures, 208–209
    model, 109–127
    per page, 97
    prerun, 34, 218–219, 232
    publishing, 33–34, 94–97,
        109–127, 197–198,
        218–221, 232–234
    reviewing, 111–112, 118–119
    runoff, 112–114, 219, 232
    size of journal, 223–225
    total, 220–221, 226–229
    trends, 226–229
    typesetting, 112–114
    use, 176–183, 217, 220–221,
        227–229

Data bases
    cost, 155–157
    directories, 150
    federal contribution, 157–158
    number, 151

searching, 153–155
size, 151–152
use, 166

Economic analysis, 10, 94–99,
    207–263
Economic growth, 24
Editing costs, 110–111, 117–118
Editorial processing center,
    69–72
Effectiveness measures, 211
Elasticity of demand, 232
Electronic processes, 290–296,
    306–311

Federal funding, 75–76, 78–79,
    125–127, 146, 157–158,
    180, 253–263
Fixed costs, 211–212, 232
Foreign authors, 200–206
Foreign subscriptions, 198
Formal versus informal
    communications, 3
Functions, 12–18, 220
Future of abstracting and
    indexing services,
    300–303

Geographic coverage, 93–94

History, 7–10

Information centers, 130–131
Information transfer
    direct versus indirect, 3
    elapsed time to publication,
        5–6
    formal versus informal, 3
    functions, 12–18
    importance, 1
    participants, 188

Journal articles
    acceptance rates, 67–68
    age when read, 172
    amount of reading, 163–165,
        193–197, 303–305
    benefits from reading,
        175–176
    cost of use, 176–183
    federal funding, 76, 78–79
    form when read, 167
    methods of identification, 165
    number, 59–62, 81–88
    number of copies transmitted,
        192–196
    page charges, 72–74, 77–79,
        250–253
    purpose of reading, 168
    rejection rates, 67–68
    reprints, 75, 222
    time spent reading, 173–175
Journals
    acceptance rates, 67–68
    advertising income, 237
    alternatives to subscriptions,
        229–230
    authors, 49–80, 218–219
    cost of use, 176–183, 217,
        220–221, 227–229
    cost per journal, 97
    cost per page, 97
    country of publication, 90–91
    definition, 1
    expenditures on, 2, 26–28
    fixed costs, 211–212, 232
    foreign subscriptions, 198
    geographic coverage, 93–94
    history, 7–10
    income, 101, 103, 106–109
    language of publication,
        90–91
    library costs, 142–146, 219
    number of articles, 25, 59–62

number of subscriptions, 30–32, 85–88, 99–101
number published, 81–88, 290–291
prerun costs, 34, 218–219, 232
prices, 32–34, 97–99, 229–232, 234–238, 243–250
public goods properties, 238
publishers, 89–91
publishing costs, 33–34, 94–97, 109–127, 197–198, 218–221, 232–234
purposes of reading, 168–170
rejection rates, 67–68
relationship with number of scientists and engineers, 85
runoff costs, 112–114, 219, 232
small, 40, 275
social benefits, 239–243
time spent reading, 2, 21, 173–175
time spent writing,
use over time, 217–218
weaknesses, 37–41
Journal system
communication paths, 187–200, 221–223
cost model, 109–127
costs by communication path, 221–223
costs by field of science, 223
costs by size of journal, 223–225
cost trends, 226–229
economics, 10, 94–99, 207–263
effects of modifications, 221–223

flow diagram, 187–188
generic functions, 12–18
libraries, 11
manuscript flow, 67–72, 192
motivation to change, 45
participants, 188
total cost, 220–221, 226–229
transmission function, 220

Language of publications, 90–91
Librarians' salaries, 227
Libraries
copyright, 38–41
cost-effectiveness, 37–142
costs, 35–37, 145–146, 219
electronics, 42–43, 46
expenditures, 132–142
federal contribution, 146
holdings, 137–139
interlibrary loan, 38, 142, 193, 199, 297–300
journal acquisitions, 143
journal cancellations, 211, 298
journal maintenance, 143
journal storage, 144
journal subscriptions, 132, 138–139, 142
journal use, 145, 195–196
National Periodicals Center, 42–43
number of U.S. libraries, 129–132
salaries, 227
Literature growth rate, 21–25

Marginal costs, 238–239
Microforms, 292
Multiattribute analysis, 231–232

National Periodicals Center, 42–44

Number
  of copies of journal articles
      transmitted, 192–196
  of journal articles, 25, 59–62,
      81–88
  of journals published, 81–88,
      290–291
  of manuscripts, 192
  of subscriptions, 30–32, 85–
      88, 99–101
  of U.S. libraries, 129–132

Optical character recognition,
      290, 307

Page charges, 72–74, 77–79,
      250–253
Performance measures,
      207–210
Periodicals
  age, 91–93
  definition, 81
  number, 90
  types of publishers, 90
Photocopying, 38–41, 143,
      276–284
Prerun costs, 34, 218–219, 232
Price-demand relationships,
      229–232
Prices, 32–34, 97–99, 229–232,
      234–238, 243–250
Probability, 71
Public goods properties, 238
Publishers
  country of publication, 90–91
  number of manuscripts, 192
  types of, 89–90
Publishing costs, 33–34, 94–97,
      197–198, 211, 218–221,
      232–234, 290–291
  cost model components,
      109–127
  editing, 110–111, 117–118

by field of science, 119
reviewing, 111–112, 118–119
runoff costs, 112–114, 219
trends, 120–125
typesetting, 112–114

Reading
  amount of reading, 162–165,
      194
  from copies transmitted from
      libraries, 195–196
  from copies transmitted by
      publishers, 194–195
  cost, 220
  source of articles, 36–37, 166
  time saved, 176
  time spent reading, 2, 21,
      173–175
Rejection rates, 67–68
Reprints, 75, 222
Reviewing costs, 111–112,
      118–119

Scientists and engineers
  geographic distribution,
      52–54
  library expenditures, 135–137
  nonreaders, 162
  number, 22–24, 49–53, 60–61
  number per journal, 85
  salaries, 77, 227
  subscription decisions,
      265–276
  time spent reading, 2, 21
  time spent writing, 21
Searching of data bases,
      153–155
Subscription decisions, 265–276
Subscriptions
  decisions regarding, 265–276
  income, 101, 103, 106–109
  per scientist or engineer, 31,
      85, 104–105

by size of journal, 88, 103
by subscriber type, 30, 100

Time spent reading, 173–175
Typesetting costs, 112–114

Usage
of abstracting and indexing
services, 34–35
of data bases, 166

of journal articles, 303–305,
163–165, 193–197
of journals in libraries, 145,
195–196
User behavior, 211
Utility theory, 64

Video discs, 293

Word processing, 41–46